Tucholsky  Wagner  Zola  Scott  Sydow  Freud  Schlegel
Turgenev  Wallace  Fonatne
Twain  Walther von der Vogelweide  Fouqué  Friedrich II. von Preußen
Weber  Freiligrath  Frey
Fechner  Fichte  Weiße Rose  von Fallersleben  Kant  Ernst  Richthofen  Frommel
Hölderlin
Engels  Fielding  Eichendorff  Tacitus  Dumas
Fehrs  Faber  Flaubert
Maximilian I. von Habsburg  Eliasberg  Ebner Eschenbach
Feuerbach  Ewald  Fock  Eliot  Zweig  Vergil
Goethe  Elisabeth von Österreich  London
Mendelssohn  Balzac  Shakespeare  Dostojewski  Ganghofer
Lichtenberg  Rathenau  Doyle  Gjellerup
Trackl  Stevenson  Hambruch
Mommsen  Tolstoi  Lenz  Hanrieder  Droste-Hülshoff
Thoma  von Arnim
Dach  Verne  Hägele  Hauff  Humboldt
Reuter  Rousseau  Hagen  Hauptmann  Gautier
Karrillon  Garschin  Defoe  Hebbel  Baudelaire
Damaschke  Descartes  Hegel  Kussmaul  Herder
Wolfram von Eschenbach  Dickens  Schopenhauer  Rilke  George
Bronner  Darwin  Melville  Grimm  Jerome  Bebel  Proust
Campe  Horváth  Aristoteles  Voltaire  Federer  Herodot
Bismarck  Vigny  Barlach  Heine
Gengenbach  Grillparzer  Georgy
Storm  Casanova  Lessing  Tersteegen  Gilm
Chamberlain  Langbein  Gryphius
Brentano  Claudius  Schiller  Lafontaine  Kralik  Iffland  Sokrates
Strachwitz  Bellamy  Schilling
Katharina II. von Rußland  Gerstäcker  Raabe  Gibbon  Tschechow
Löns  Hesse  Hoffmann  Gogol  Wilde  Gleim  Vulpius
Luther  Heym  Hofmannsthal  Klee  Hölty  Morgenstern  Goedicke
Roth  Heyse  Klopstock  Kleist
Luxemburg  Puschkin  Homer  Mörike
La Roche  Horaz  Musil
Machiavelli  Kierkegaard  Kraft  Kraus
Navarra  Aurel  Musset  Kind  Moltke
Nestroy  Marie de France  Lamprecht  Kirchhoff  Hugo
Nietzsche  Nansen  Laotse  Ipsen  Liebknecht
Marx  Lassalle  Gorki  Klett  Leibniz  Ringelnatz
von Ossietzky  May  vom Stein  Lawrence  Irving
Petalozzi  Platon  Knigge
Sachs  Pückler  Michelangelo  Kafka
Poe  Liebermann  Kock
de Sade  Praetorius  Mistral  Zetkin  Korolenko

The publishing house tredition has created the series **TREDITION CLASSICS**. It contains classical literature works from over two thousand years. Most of these titles have been out of print and off the bookstore shelves for decades.

The book series is intended to preserve the cultural legacy and to promote the timeless works of classical literature. As a reader of a **TREDITION CLASSICS** book, the reader supports the mission to save many of the amazing works of world literature from oblivion.

The symbol of **TREDITION CLASSICS** is Johannes Gutenberg (1400 – 1468), the inventor of movable type printing.

With the series, tredition intends to make thousands of international literature classics available in printed format again – worldwide.

All books are available at book retailers worldwide in paperback and in hardcover. For more information please visit: www.tredition.com

tredition was established in 2006 by Sandra Latusseck and Soenke Schulz. Based in Hamburg, Germany, tredition offers publishing solutions to authors and publishing houses, combined with worldwide distribution of printed and digital book content. tredition is uniquely positioned to enable authors and publishing houses to create books on their own terms and without conventional manufacturing risks.

For more information please visit: www.tredition.com

# British Airships, Past, Present, and Future

George Whale

# Imprint

This book is part of the TREDITION CLASSICS series.

Author: George Whale
Cover design: toepferschumann, Berlin (Germany)

Publisher: tredition GmbH, Hamburg (Germany)
ISBN: 978-3-8495-0724-4

www.tredition.com
www.tredition.de

Copyright:
The content of this book is sourced from the public domain.

The intention of the TREDITION CLASSICS series is to make world literature in the public domain available in printed format. Literary enthusiasts and organizations worldwide have scanned and digitally edited the original texts. tredition has subsequently formatted and redesigned the content into a modern reading layout. Therefore, we cannot guarantee the exact reproduction of the original format of a particular historic edition. Please also note that no modifications have been made to the spelling, therefore it may differ from the orthography used today.

# Contents

CHAPTER I
INTRODUCTION

CHAPTER II
EARLY AIRSHIPS AND THEIR DEVELOPMENT
TO THE PRESENT DAY

CHAPTER III
BRITISH AIRSHIPS BUILT BY PRIVATE FIRMS

CHAPTER IV
BRITISH ARMY AIRSHIPS

CHAPTER V
EARLY DAYS OF THE NAVAL AIRSHIP SECTION--
PARSEVAL AIRSHIPS,
ASTRA-TORRES TYPE, ETC.

CHAPTER VI
NAVAL AIRSHIPS: THE NON-RIGIDS--
    S.S. TYPE
    COASTAL AND C STAR AIRSHIPS
    THE NORTH SEA AIRSHIP

CHAPTER VII
NAVAL AIRSHIPS: THE RIGIDS
    RIGID AIRSHIP NO. 1
    RIGID AIRSHIP NO. 9
    RIGID AIRSHIP NO. 23 CLASS
    RIGID AIRSHIP NO. 23 X CLASS
    RIGID AIRSHIP NO. 31 CLASS
    RIGID AIRSHIP NO. 33 CLASS

CHAPTER VIII
THE WORK OF THE AIRSHIP IN THE WORLD WAR

CHAPTER IX
THE FUTURE OF AIRSHIPS

# CHAPTER I

## INTRODUCTION

Lighter-than-air craft consist of three distinct types: Airships, which are by far the most important, Free Balloons, and Kite Balloons, which are attached to the ground or to a ship by a cable. They derive their appellation from the fact that when charged with hydrogen, or some other form of gas, they are lighter than the air which they displace. Of these three types the free balloon is by far the oldest and the simplest, but it is entirely at the mercy of the wind and other elements, and cannot be controlled for direction, but must drift whithersoever the wind or air currents take it. On the other hand, the airship, being provided with engines to propel it through the air, and with rudders and elevators to control it for direction and height, can be steered in whatever direction is desired, and voyages can be made from one place to another--always provided that the force of the wind is not sufficiently strong to overcome the power of the engines. The airship is, therefore, nothing else than a dirigible balloon, for the engines and other weights connected with the structure are supported in the air by an envelope or balloon, or a series of such chambers, according to design, filled with hydrogen or gas of some other nature.

It is not proposed, in this book, to embark upon a lengthy and highly technical dissertation on aerostatics, although it is an intricate science which must be thoroughly grasped by anyone who wishes to possess a full knowledge of airships and the various problems which occur in their design. Certain technical expressions and terms are, however, bound to occur, even in the most rudimentary work on airships, and the main principles underlying airship construction will be described as briefly and as simply as is possible.

The term "lift" will appear many times in the following pages, and it is necessary to understand what it really means. The difference between the weight of air displaced and the weight of gas in a balloon or airship is called the "gross lift." The term "disposable," or "nett" lift, is obtained by deducting the weight of the structure, cars, machinery and other fixed weights from the gross lift. The resultant weight obtained by this calculation determines the crew, ballast, fuel and other necessities which can be carried by the balloon or airship.

The amount of air displaced by an airship can be accurately weighed, and varies according to barometric pressure and the temperature; but for the purposes of this example we may take it that under normal conditions air weighs 75 lb. per 1,000 cubic feet. Therefore, if a balloon of 1,000 cubic feet volume is charged with air, this air contained will weigh 75 lb. It is then manifest that a balloon filled with air would not lift, because the air is not displaced with a lighter gas.

Hydrogen is the lightest gas known to science, and is used in airships to displace the air and raise them from the ground. Hydrogen weighs about one-fifteenth as much as air, and under normal conditions 1,000 cubic feet weighs 5 lb. Pursuing our analogy, if we fill our balloon of 1,000 cubic feet with hydrogen we find the gross lift is as follows:

> 1,000 cubic feet of air weighs 75 lb. 1,000 cubic feet of hydrogen weighs 5 lb. ------ The balance is the gross lift of the balloon 70 lb.

It follows, then, that apart from the weight of the structure itself the balloon is 70 lb. lighter than the air it displaces, and provided that it weighs less than 70 lb. it will ascend into the air.

As the balloon or airship ascends the density of the air decreases as the height is increased. As an illustration of this the barometer falls, as everyone knows, the higher it is taken, and it is accurate to say that up to an elevation of 10,000 feet it falls one inch for every 1,000 feet rise. It follows that as the pressure of the air decreases, the volume of the gas contained expands at a corresponding rate. It has been shown that a balloon filled with 1,000 feet of hydrogen has a lift of 70 lb. under normal conditions, that is to say, at a barometric pressure of 80 inches. Taking the barometric pressure at 2 inches lower, namely 28, we get the following figures:

> 1,000 cubic feet of air weighs 70 lb. 1,000 cubic feet of hydrogen weighs 4.67 " --------- 65.33 lb.

It is therefore seen that the very considerable loss of lift, 4.67 lb. per 1,000 cubic feet, takes place with the barometric pressure 2 inches lower, from which it may be taken approximately that 1/30 of the volume gross lift and weight is lost for every 1,000 feet rise. From this example it is obvious that the greater the pressure of the atmosphere, as indicated by the barometer, the greater will be the lift of the airship or balloon.

Temperature is another factor which must be considered while discussing lift. The volume of gas is affected by temperature, as gases expand or contract about 1/500 part for every degree Fahrenheit rise or fall in temperature.

In the case of the 1,000 cubic feet balloon, the air at 30 inches barometric pressure and 60 degrees Fahrenheit weighs 75 lb., and the hydrogen weighs 5 lb.

At the same pressure, but with the temperature increased to 90 degrees Fahrenheit, the air will be expanded and 1,000 cubic feet of air will weigh only 70.9 lb., while 1,000 cubic feet of hydrogen will weigh 4.7 lb.

The lift being the difference between the weight of the volume of air and the weight of the hydrogen contained in the balloon, it will be seen that with the temperature at 60 degrees Fahrenheit the lift is 75 lb. - 5 lb. = 70 lb., while the temperature, having risen to 90 degrees, the lift now becomes 70.9 lb. - 4.7 lb. = 66.2 lb.

Conversely, with a fall in the temperature the lift is increased.

We accordingly find from the foregoing observations that at the start of a voyage the lift of an airship may be expected to be greater when the temperature is colder, and the greater the barometric pressure so will also the lift be greater. To put this into other words, the most favourable conditions for the lift of an airship are when the weather is cold and the barometer is high.

It must be mentioned that the air and hydrogen are not subject in the same way to changes of temperature. Important variations in lift may occur when the temperature of the gas inside the envelope becomes higher, owing to the action of the sun, than the air which surrounds it. A difference of some 20 degrees Fahrenheit may result between the gas and the air temperatures; this renders it highly necessary that the pilot should by able to tell at any moment the relative temperatures of gas and air, as otherwise a false impression will be gained of the lifting capacity of the airship.

The lift of an airship is also affected by flying through snow and rain. A considerable amount of moisture can be taken up by the fabric and suspensions of a large airship which, however, may be largely neutralized by the waterproofing of the envelope. Snow, as a rule, is brushed off the surface by the passage of the ship through the air, though in the event of its freezing suddenly, while in a melting state, a very considerable addition of weight might be caused. There have been many instances of airships flying through snow, and as far as is known no serious difficulty has been encountered through the adhesion of this substance. The humidity of the air may also cause slight variations in lift, but for rough calculations it may be ignored, as the difference in lift is not likely to amount to more than 0.3 lb. per 1,000 cubic feet of gas.

The purity of hydrogen has an important effect upon the lift of an airship. One of the greatest difficulties to be contended with is maintaining the hydrogen pure in the envelope or gasbags for any length of time. Owing to diffusion gas escapes with extraordinary rapidity, and if the fabric used is not absolutely gastight the air finds its way in where the gas has escaped. The maximum purity of gas in an airship never exceeds 98 per cent by volume, and the following example shows how greatly lift can be reduced:

Under mean atmospheric conditions, which are taken at a temperature of 55 degrees Fahrenheit, and the barometer at 29.5 inches, the lift of 1,000 cubic feet of hydrogen at 98 per cent purity is 69.6 lb. Under same conditions at 80 per cent purity the lift of 1,000 cubic feet of hydrogen is 56.9 lb., a resultant loss of 12.9 lb. per 1,000 cubic feet.

The whole of this statement on "lift" can now be condensed into three absolute laws:

1. Lift is directly proportional to barometric pressure.

2. Lift is inversely proportional to absolute temperature.

3. Lift is directly proportional to purity.

# AIRSHIP DESIGN

The design of airships has been developed under three distinct types, the Rigid, the Semi-Rigid, and the Non-Rigid.

The rigid, of which the German Zeppelin is the leading example, consists of a framework, or hull composed of aluminium, wood, or other materials from which are suspended the cars, machinery and other weights, and which of itself is sufficiently strong to support its own weight. Enclosed within this structure are a number of gas chambers or bags filled with hydrogen, which provide the necessary buoyancy. The hull is completely encased within a fabric outer cover to protect the hull framework and bags from the effects of weather, and also to temper the rays of the sun.

The semi-rigid, which has been exploited principally by the Italians with their Forlanini airships, and in France by Lebaudy, has an envelope, in some cases divided into separate compartments, to which is attached close underneath a long girder or keel. This supports the car and other weights and prevents the whole ship from buckling in the event of losing gas. The semi-rigid type has been practically undeveloped in this country.

The non-rigid, of which we may now claim to be the leading builders, is of many varieties, and has been developed in several countries. In Germany the chief production has been that of Major von Parseval, and of which one ship was purchased by the Navy shortly before the outbreak of war. In the earliest examples of this type the car was slung a long way from the envelope and was supported by wires from all parts. This necessitated a lofty shed for its accommodation as the ship was of great overall height; but this difficulty was overcome by the employment of the elliptical and trajectory bands, and is described in the chapter dealing with No. 4.

A second system is that of the Astra-Torres. This envelope is trilobe in section, with internal rigging, which enables the car to be slung very close up to the envelope. The inventor of these envelopes was a Spaniard, Senor Torres Quevedo, who manufactured them in conjunction with the Astra Company in Paris. This type of envelope has been employed in this country in the Coastal, C Star, and North

Sea airships, and has been found on the whole to give good results. It is questionable if an envelope of streamline shape would not be easier to handle, both in the air and on the landing ground, and at present there are partisans of both types.

Thirdly, there is the streamline envelope with tangential suspensions, which has been adopted for all classes of the S.S. airship, and which has proved for its purpose in every way highly satisfactory.

Of these three types the rigid has the inherent disadvantage of not being able to be dismantled, if it should become compelled to make a forced landing away from its base. Even if it were so fortunate as to escape damage in the actual landing, there is the practical certainty that it would be completely wrecked immediately any increase occurred in the force of the wind. On the other hand, for military purposes, it possesses the advantage of having several gas compartments, and is in consequence less susceptible to damage from shell fire and other causes.

Both the semi-rigid and the non-rigid have the very great advantage of being easily deflated and packed up. In addition to the valves, these ships have a ripping panel incorporated in the envelope which can easily be torn away and allows the gas to escape with considerable rapidity. Innumerable instances have occurred of ships being compelled to land in out-of-the-way places owing to engine failure or other reasons; they have been ripped and deflated and brought back to the station without incurring any but the most trifling damage.

Experience in the war has proved that for military purposes the large rigid, capable of long hours of endurances and the small non-rigid made thoroughly reliable, are the most valuable types for future development. The larger non-rigids, with the possible exception of the North Sea, do not appear to be likely to fulfil any very useful function.

Airship design introduces so many problems which are not met with in the ordinary theory of structures, that a whole volume could easily be devoted to the subject, and even then much valuable information would have to be omitted from lack of space. It is, therefore, impossible, in only a section of a chapter, to do more than indicate in the briefest manner a few salient features concerning these problems. The suspension of weights from the lightest possible gas compartment must be based on the ordinary principles of calculating the distribution loads as in ships and other structures. In the non-rigid, the envelope being made of flexible fabric has, in itself, no rigidity whatsoever, and its shape must be maintained by the internal pressure kept slightly in excess of the pressure outside. Fabric is capable of resisting tension, but is naturally not able to resist compression. If the car was rigged beneath the centre of the envelope with vertical suspensions it would tend to produce compression in the underside of the envelope, owing to the load not being fully distributed. This would cause, in practice, the centre portion of the envelope to sag downwards, while the ends would have a tendency to rise. The principle which has been found to be most satisfactory is to fix the points of suspension distributed over the greatest length of envelope possible proportional to the lift of gas at each section thus formed. From these points the wires are led to the car. If the car is placed close to the envelope it will be seen that the suspensions of necessity lie at a very flat angle and exert a serious longitudinal compression. This must be resisted by a high internal pressure, which demands a stouter fabric for the envelope and, therefore, increased weight. It follows that the tendency of the envelope to deform is decreased as the distance of the car from the gas compartment is increased.

One method of overcoming this difficulty is found by using the Astra-Torres design. As will be seen from the diagram of the North Sea airship, the loads are excellently distributed by the several fans of internal rigging, while external head resistance is reduced to a minimum, as the car can be slung close underneath the envelope. Moreover, the direct longitudinal compression due to the rigging is applied to a point considerably above the axis of the ship. In a large non-rigid many of these difficulties can be overcome by distributing the weight into separate cars along the envelope itself.

We have seen that as an airship rises the gas contained in the envelope expands. If the envelope were hermetically sealed, the higher the ship rose the greater would become the internal pressure, until the envelope finally burst. To avoid this difficulty in a balloon, a valve is provided through which the gas can escape. In a balloon, therefore, which ascends from the ground full, gas is lost throughout its upward journey, and when it comes down again it is partially empty or flabby. This would be an impossible situation in the case of the airship, for she would become unmanageable, owing to the buckling of the envelope and the sagging of the planes. Ballonets are therefore fitted to prevent this happening.

Ballonets are internal balloons or air compartments fitted inside the main envelope, and were originally filled with air by a blower driven either by the main engines or an auxiliary motor. These blowers were a continual source of trouble, and at the present day it has been arranged to collect air from the slip-stream of the propeller through a metal air scoop or blower-pipe and discharge it into an air duct which distributes it to the ballonets.

The following example will explain their functions:

An airship ascends from the ground full to 1,000 feet. The ballonets are empty, and remain so throughout the ascent. By the time the airship reaches 1,000 feet it will have lost 1/30th of its volume of gas which will have escaped through the valves. If the ship has a capacity of 300,000 cubic feet it will have lost 10,000 cubic feet of gas. The airship now commences to descend; as it descends the gas within contracts and air is blown into the ballonets. By the time the ground is reached 10,000 cubic feet of air will have been blown into the ballonets and the airship will have retained its shape and not be flabby.

On making a second ascent, as the airship rises the air must be let out of the ballonet instead of gas from the envelope, and by the time 1,000 feet is reached the ballonets will be empty. To ensure that this is always done the ballonet valves are set to open at less pressure than the gas valves.

It therefore follows in the example under consideration that it will not be necessary to lose gas during flight, provided that an ascent is not made over 1,000 feet.

Valves are provided to prevent the pressure in the envelope from exceeding a certain determined maximum and are fitted both to ballonets and the gaschamber. They are automatic in action, and, as we have said, the gas valve is set to blow off at a pressure in excess of that for the air valve.

In rigid airships ballonets are not provided for the gasbags, and as a consequence a long flight results in a considerable expenditure of gas. If great heights are required to be reached, it is obvious that the wastage of gas would be enormous, and it is understood that the Germans on starting for a raid on England, where the highest altitudes were necessary, commenced the flight with the gasbags only about 60 per cent full.

To stabilize the ship in flight, fins or planes are fitted to the after end of the envelope or hull. Without the horizontal planes the ship will continually pitch up and down, and without the vertical planes it will be found impossible to keep the ship on a straight course. The planes are composed of a framework covered with fabric and are attached to the envelope by means of stay wires fixed to suitable points, in the case of non-rigid ships skids being employed to prevent the edge of the plane forcing its way through the surface of the fabric. The rudder and elevator flaps in modern practice are hinged to the after edges of the planes.

The airship car contains all instruments and controls required for navigating the ship and also provides a housing for the engines. In the early days swivelling propellers were considered a great adjunct, as with their upward and downward thrust they proved of great value in landing. Nowadays, owing to greater experience, landing does not possess the same difficulty as in the past, and swivelling propellers have been abandoned except in rigid airships, and even in the later types of these they have been dispensed with.

Owing to the great range of an airship a thoroughly reliable engine is a paramount necessity. The main requirements are--firstly, that it must be capable of running for long periods without a breakdown; secondly, that it must be so arranged that minor repairs can be effected in the air; and thirdly, that economy of oil and fuel is of far greater importance to an airship than the initial weight of the engine itself.

## HANDLING AND FLYING OF AIRSHIPS

The arrangements made for handling airships on the ground and while landing, and also for moving them in the open, provide scope for great ingenuity. An airship when about to land is brought over the aerodrome and is "ballasted up" so that she becomes considerably lighter than the air which she displaces. The handling party needs considerable training, as in gusty weather the safety of the ship depends to a great extent upon its skill in handling her. The ship approaches the handling party head to wind and the trail rope is dropped; it is taken by the handling party and led through a block secured to the ground and the ship is slowly hauled down. When near the ground the handling party seize the guys which are attached to the ship at suitable points, other detachments also support the car or cars, as the case may be, and the ship can then be taken into the shed.

In the case of large airships the size of the handling party has to be increased and mechanical traction is also at times employed.

As long as the airship is kept head to wind, handling on the ground presents little difficulty; on many occasions, however, unless the shed is revolving, as is the case on certain stations in Germany, the wind will be found to be blowing across the entrance to the shed. The ship will then have to be turned, and during this operation, unless great discretion is used, serious trouble may be experienced.

Many experiments have been and are still being conducted to determine the best method of mooring airships in the open. These will be described and discussed at some length in the chapter devoted to the airship of the future.

During flight certain details require attention, and carelessness on the pilot's part, even on the calmest of days, may lead to disaster. The valves and especially the gas valves should be continually tested, as on occasions they have been known to jam, and the loss of gas has not been discovered until the ship had become unduly heavy.

Pressure should be kept as constant as possible. Most airships work up to 30 millimetres as a maximum and 15 millimetres as a minimum flying pressure. During a descent the pressure should be watched continuously, as it may fall so low as to cause the nose to blow in. This will right itself when the speed is reduced or the pressure is raised, but there is always the danger of the envelope becoming punctured by the bow stiffeners when this occurs.

## HOUSING ACCOMMODATION FOR AIRSHIPS, ETC.

During the early days of the war, when stations were being equipped, the small type of airship was the only one we possessed. The sheds to accommodate them were constructed of wood both for cheapness and speed of construction and erection. These early sheds were all of very similar design, and were composed of trestles with some ordinary form of roof-truss. They were covered externally with corrugated sheeting. The doors have always been a source of difficulty, as they are compelled to open for the full width of the shed and have to stand alone without support. They are fitted with wheels which run on guide rails, and are opened by means of winches and winding gear.

The later sheds built to accommodate the rigid airship are of much greater dimensions, and are constructed of steel, but otherwise are of much the same design.

The sheds are always constructed with sliding doors at either end, to enable the ship to be taken out of the lee end according to the direction of the wind.

It has been the practice in this country to erect windscreens in order to break the force of the wind at the mouth of the shed. These screens are covered with corrugated sheeting, but it is a debatable point as to whether the comparative shelter found at the actual opening of the shed is compensated for by the eddies and air currents which are found between the screens themselves. Experiments have been carried out to reduce these disturbances, in some cases by removing alternate bays of the sheeting and in other cases by substituting expanded metal for the original corrugated sheets.

It must be acknowledged that where this has been done, the airships have been found easier to handle.

At the outbreak of war, with the exception of a silicol plant at Kingsnorth, now of obsolete type, and a small electrolytic plant at Farnborough, there was no facility for the production of hydrogen in this country for the airship service.

When the new stations were being equipped, small portable silicol plants were supplied capable of a small output of hydrogen. These were replaced at a later date by larger plants of a fixed type, and a permanent gas plant, complete with gasholders and high pressure storage tanks was erected at each station, the capacity being 5,000 or 10,000 cubic feet per hour according to the needs of the station.

With the development of the rigid building programme, and the consequent large requirements of gas, it was necessary to reconsider the whole hydrogen situation, and after preliminary experimental work it was decided to adopt the water gas contact process, and plants of this kind with a large capacity of production were erected at most of the larger stations. At others electrolytic plants were put down. Hydrogen was also found to be the bye-product of certain industries, and considerable supplies were obtained from commercial firms, the hydrogen being compressed into steel cylinders and dispatched to the various stations.

Before concluding this chapter, certain words must be written on parachutes. A considerable controversy raged in the press and elsewhere a few months before the cessation of hostilities on the subject of equipping the aeroplane with parachutes as a life-saving device. In the airship service this had been done for two years. The best type of parachute available was selected, and these were fitted according to circumstances in each type of ship. The usual method is to insert the parachute, properly folded for use, in a containing case which is fastened either in the car or on the side of the envelope as is most convenient. In a small ship the crew are all the time attached to their parachutes and in the event of the ship catching fire have only to jump overboard and possess an excellent chance of being saved. In rigid airships where members of the crew have to move from one end of the ship to the other, the harness is worn and parachutes are disposed in the keel and cars as are lifebuoys in seagoing vessels. Should an emergency arise, the nearest parachute can be attached to the harness by means of a spring hook, which is the work of a second, and a descent can be made.

It is worthy of note that there has never been a fatal accident or any case of a parachute failing to open properly with a man attached.

The material embodied in this chapter, brief and inadequate as it is, should enable the process of the development of the airship to be easily followed. Much has been omitted that ought by right to have been included, but, on the other hand, intricate calculations are apt to be tedious except to mathematicians, and these have been avoided as far as possible in the following pages.

# CHAPTER II

## EARLY AIRSHIPS AND THEIR DEVELOPMENT TO THE PRESENT DAY

The science of ballooning had reached quite an advanced stage by the middle of the eighteenth century, but the construction of an airship was at that time beyond the range of possibility. Discussions had taken place at various times as to the practicability of rendering a balloon navigable, but no attempts had been made to put these points of argument to a practical test.

Airship history may be said to date from January 24th, 1784. On that day Brisson, a member of the Academy in Paris, read before that Society a paper on airships and the methods to be utilized in propelling them. He stated that the balloon, or envelope as it is now called, must be cylindrical in shape with conical ends, the ratio of diameter to length should be one to five or one to six and that the smallest cross-sectional area should face the wind. He proposed that the method of propulsion should be by oars, although he appeared to be by no means sanguine if human strength would be sufficient to move them. Finally, he referred to the use of different currents of the atmosphere lying one above the other.

This paper caused a great amount of interest to be taken in aeronautics, with the result that various Frenchmen turned their attention to airship design and production. To France must be due the acknowledgment that she was the pioneer in airship construction and to her belongs the chief credit for early experiments.

At a later date Germany entered the lists and tackled the problems presented with that thoroughness so characteristic of the nation. It is just twenty-one years ago since Count Zeppelin, regardless of public ridicule, commenced building his rigid airships, and in that time such enormous strides were made that Germany, at the outbreak of the war, was ahead of any other country in building the large airship.

In 1908 Italy joined the pioneers, and as regards the semi-rigid is in that type still pre-eminent. Great Britain, it is rather sad to say, adopted the policy of "wait and see," and, with the exception of a few small ships described in the two succeeding chapters, had produced nothing worthy of mention before the outbreak of the great European war. She then bestirred herself, and we shall see later that she has produced the largest fleet of airships built by any country and, while pre-eminent with the non-rigid, is seriously challenging Germany for the right to say that she has now built the finest rigid airship.

# FRANCE

To revert to early history, in the same year in which Brisson read his paper before the Academy, the Duke of Chartres gave the order for an airship to the brothers Robert, who were mechanics in Paris. This ship was shaped like a fish, on the supposition that an airship would swim through the air like a fish through water. The gas-chamber was provided with a double envelope, in order that it might travel for a long distance without loss of gas.

The airship was built in St. Cloud Park; in length it was 52 feet with a diameter of 82 feet, and was ellipsoidal in shape with a capacity of 30,000 cubic feet. Oars were provided to propel it through the air, experiments having proved that with two oars of six feet diameter a back pressure of 90 lb. was obtained and with four oars 140 lb.

On July 6th in the same year the first ascent was made from St. Cloud. The passengers were the Duke of Chartres, the two brothers Robert and Colin-Hulin. No valves having been fitted, there was no outlet for the expansion of gas and the envelope was on the point of bursting, when the Duke of Chartres, with great presence of mind, seized a pole and forced an opening through both the envelopes. The ship descended in the Park of Meudon.

On September 19th the airship made a second ascent with the same passengers as before, with the exception of the Duke. According to the report of the brothers Robert, they succeeded in completing an ellipse and then travelled further in the direction of the wind without using the oars or steering arrangements. They then deviated their course somewhat by the use of these implements and landed at Bethune, about 180 miles distant from Paris.

In those days it was considered possible that a balloon could be rendered navigable by oars, wings, millwheels, etc., and it was not until the last decades of the nineteenth century, when light and powerful motors had been constructed, that the problem became really practical of solution.

During the nineteenth century several airships were built in France and innumerable experiments were carried out, but the vessels produced were of little real value except in so far as they stimulated their designers to make further efforts. Two of these only will be mentioned, and that because the illustrations show how totally different they were from the airship of to-day.

In 1834 the Compte de Lennox built an airship of 98,700 cubic feet capacity. It was cylindrical in form with conical ends, and is of interest because a small balloon or ballonet, 7,050 cubic feet contents, was placed inside the larger one for an air filling. A car 66 feet in length was rigged beneath the envelope by means of ropes eighteen inches long. Above the car the envelope was provided with a long air cushion in connection with a valve. The intention was by compression of the air in the cushion and the inner balloon, to alter the height of the airship, in order to travel with the most favourable air currents. The motive power was 20 oar propellers worked by men.

This airship proved to be too heavy on completion to lift its own weight, and was destroyed by the onlookers.

The next airship, the Dupuy de Lome, is of interest because the experiments were carried out at the cost of the State by the French Government. This ship consisted of a spindle-shaped balloon with a length of 112 feet, diameter of 48 1/2 feet and a volume of 121,800 cubic feet. An inner air balloon of 6,000 cubic feet volume was contained in the envelope. The method of suspension was by means of diagonal ropes with a net covering. A rudder in the form of a triangular sail was fitted beneath the envelope and at the after part of the ship. The motive power was double-winged screws 29 feet 6 inches diameter, to be worked by four to eight men.

On her trials the ship became practically a free balloon, an independent velocity of about six miles per hour being achieved and deviation from the direction of the wind of ten degrees.

At the close of the nineteenth century Santos-Dumont turned his attention to airships. The experiments which he carried out marked a new epoch and there arose the nucleus of the airship as we know it to-day. Between the years 1898 and 1905 he had in all built fourteen airships, and they were continually improved as each succeeding one made its appearance. In the last one he made a circular flight; starting from the aerodrome of the aero club, he flew round the Eiffel Tower and back to the starting point in thirty-one minutes on October 19th, 1902. For this feat the Deutsch prize was awarded to him.

The envelopes he used were in design much nearer approach to a streamline form than those previously adopted, but tapered to an extremely fine point both at the both and stem. For rigging he employed a long nacelle, in the centre of which was supported the car, and unusually long suspensions distributed the weight throughout practically the entire length of the envelope. To the name of Santos-Dumont much credit is due. He may be regarded as the originator of the airship for pleasure purposes, and by his success did much to popularize them. He also was responsible to a large extent for the development and expansion of the airship industry in Paris.

At a little later date, in 1902 to be precise, the Lebaudy brothers, in conjunction with Julliot, an engineer, and Surcoup, an aeronaut, commenced building an airship of a new type. This ship was a semirigid and was of a new shape, the envelope resembling in external appearance a cigar. In length it was 178 feet with a diameter of 30 feet and the total capacity was 64,800 cubic feet. This envelope was attached to a rigid elliptical keel-shaped girder made of steel tubes, which was about a third of the length of the ship. The girder was covered with a shirting and intended to prevent the ship pitching and rolling while in flight. A horizontal rudder was attached to the under side of this girder, while right aft a large vertical rudder was fixed.

A small car was suspended by steel rods at a distance of 17 feet 9 inches from the girder, with a framework built up underneath to absorb the shock on landing.

A 35 horse-power Daimler-Mercedes motor, weighing some 800 lb. without cooling water and fuel, drove two twin-bladed propellers on either side of the car.

In the year 1903 a number of experimental flights were made with this ship and various details in the construction were continually introduced. The longest flight was 2 hours 46 minutes. Towards the end of that year, while a voyage was being made from Paris to Chalais Meudon, the airship came in contact with a tree and the envelope was badly torn.

In the following year it was rebuilt, and the volume was slightly increased with fixed and movable planes added to increase the stability. After several trips had been made, the airship again on landing came in contact with a tree and was burst.

The ship was rebuilt and after carrying out trials was purchased by the French Army. The Lebaudy airship had at that time been a distinct success, and in 1910 one was purchased for the British Government by the readers of the Morning Post.

In the ten-ton Lebaudy the length of the keel framework was greatly extended, and ran for very nearly the full length of the envelope. The disadvantage of this ship was its slowness, considering its size and power, and was due to the enormous resistance offered by the framework and rigging.

Airships known as the "Clement-Bayard" were also built about this time. They were manufactured by the Astra Company in conjunction with Monsieur Clement, a motor engineer.

In later days vessels were built by the Astra Company of the peculiar design introduced by Senor Torres. These ships, some of which were of considerable size, were highly successful, and we became purchasers at a later date of several.

The Zodiac Company also constructed a number of small ships which were utilized during the war for anti-submarine patrol. It cannot be said, however, that the French have fulfilled their early promise as airship designers, the chief reason for this being that the airship is peculiarly suitable for work at sea and the French relied on us to maintain the commerce routes on the high seas and concentrated their main efforts on defeating the Germans in the field, in which as all the world acknowledges they were singularly successful and hold us under an eternal obligation.

## GERMANY

The progress and development of the airship in Germany must now be considered; it will be seen that, although the production of satisfactory ships was in very few hands, considerable success attended their efforts in the early days of the twentieth century.

In 1812, Leppig built an airship at the cost of the State at Woronzowo in Russia. This was of the shape of a fish with a rigid framework beginning at the height of the longitudinal axis.

The lower keel-shaped part of the same formed the car. Two fans were attached to the sides and a tail piece was provided behind to act as a rudder. The ship was inflated, but structural damage occurred during this operation and rendered it incapable of flight.

In 1836, Georg Rebenstein, of Nurnburg, was considering the use of the fall of inclined planes to obtain horizontal motion.

Nothing of importance was produced until a much later date, when in 1885 M. Wolf constructed an envelope of 26,500 cubic feet. An engine and propeller were fixed in a triangular framework in front of the airship, supported by the steam pipe of a steam engine fixed under the body of the envelope. The framework lacked rigidity, and the envelope tore during inflation and the airship failed to ascend.

In the following year Dr. Woelfert, of Berlin, produced a cigar-shaped envelope, to which was attached rigidly a long bamboo framework containing the car. An 8 horse-power benzine Daimler motor drove a twin-bladed aluminium propeller, and another propeller for vertical movement was provided beneath the car. Four trial flights were attempted, but on each occasion the motor gave unsatisfactory results, and Woelfert sought to improve it with a benzine vaporizer of his own pattern. This improvement was not a success, as during the last flight an explosion took place and both Woelfert and an aeronaut named Knabe, who was accompanying him, were killed.

In 1906, Major von Parseval experimented, in Berlin, with a non-rigid type of airship. His first ship had a volume of 65,200 cubic feet, but owing to his system of suspensions, the car hung 27 feet 6 inches below the envelope. A Daimler engine was used, driving a four-bladed propeller. Owing to the great overall height of this ship, experiments were made to determine a system of rigging, enabling the car to be slung closer to the envelope, and in later types the elliptical rigging girdle was adopted. His later ships were of large dimensions and proved very satisfactory. About the same time Major Gross also built airships for the German aeronautical battalion.

It is, however, the rigid airship that has made Germany famous, and we must now glance at the evolution of these ships with which we became so familiar during the war.

The first rigid airship bearing any resemblance to those of the present day was designed by David Schwartz, and was built in St. Petersburg in 1893. It was composed of aluminium plates riveted to an aluminium framework. On inflation, the frame-work collapsed and the ship was unusable.

In 1895 he designed a second rigid airship, which was built in Berlin by Messrs. Weisspfennig and Watzesch. The hull framework was composed of aluminium and was 155 feet long, elliptical in cross section, giving a volume of 130,500 cubic feet. It was pointed in front and rounded off aft. The car, also constructed of the same material, was rigidly attached to the hull by a lattice framework, and the whole hull structure was covered in with aluminium sheeting. A 12 horse-power Daimler benzine motor was installed in the car, driving through the medium of a belt twin aluminium screw propellers; no rudders were supplied, the steering being arranged by means of a steering screw placed centrally to the ship above the top of the car. Inflation took place at the end of 1897 by a method of pressing out air-filled fabric cells which were previously introduced into the hull. This operation took three and a half hours. On the day of the first flight trials there was a fresh wind of about 17 miles per hour. The airship ascended into the air, but, apparently, could make little headway against the wind. During the trip the driving-belt became disengaged from the propellers and the ship drifted at the mercy of the wind, but sustained little damage on landing. After being deflated, the hull began to break up under the pressure of the wind and was completely destroyed by the vandalism of the spectators.

In 1898 Graf F. von Zeppelin, inspired by the example of Schwartz, and assisted by the engineers Kober and Kubler, conceived the idea of constructing a rigid airship of considerable dimensions. For this purpose a floating shed was built on Lake Constance, near to Friedrichshafen. The hull was built of aluminium lattice-work girders, and had the form of a prism of twenty-four surfaces with arch-shaped ends. In length it was 420 feet, with a diameter of 38 feet 6 inches, and its capacity was 400,000 cubic feet. The longitudinal framework was divided by a series of rings, called transverse frames, into seventeen compartments containing fabric gasbags. The transverse frames were fitted with steel wire bracings, both radial and chord, and to strengthen the whole a triangular aluminium keel of lattice work was used. A vertical and horizontal rudder were fitted to the forward portion of the ship, and aft another vertical rudder. The whole exterior of the ship was fitted with a fabric outer cover.

Two aluminium cars, each about 20 feet long, were rigidly attached to the framework of the hull. Each car was furnished with a 16 horse-power Daimler engine, driving two four-bladed screw propellers of aluminium sheeting. These propellers were situated on the side of the hull at the centre of resistance. The transmission was supplied by steel tubes with universal cross joints through the medium of bevel gears. Reversible driving arrangements were installed in the cars in order that the ship could be driven backwards and forwards. Electric bells, telegraphs, and speaking tubes were also fitted, and it can be seen that for general arrangements this airship was a long way ahead of any built at that date.

The first flight was made on July 2nd, 1900. The ship attained a speed of 17 per hour, and the numerous technical details stood the tests well. The stability was considered sufficient, and the height of flight could be altered by the horizontal rudder. The landing on the water was accomplished without difficulty, and could be regarded as free from danger. The faults requiring remedy were, firstly, the upper cross stays, which buckled in flight owing to insufficient strength for the length of the hull; secondly, the gasbags were not sufficiently gastight and, thirdly, the power of the engines were not sufficient for such a heavy ship.

This airship was broken up in 1902.

In 1905 the second ship of the series was completed. She was of nearly the same size as the previous ship, but the workmanship was much superior. Increased engine-power was also supplied, as in this instance two 85 horse-power Mercedes engines were fitted. This ship was destroyed by a storm while landing during the next year.

The third ship, which was completed in 1906, was the first Zeppelin airship acquired by the Government, and lasted for a considerable time, being rebuilt twice, first in 1908 and again in 1911. She was slightly larger than the previous two.

The building was continued, and up to the outbreak of war no fewer than twenty-five had been completed. It is impossible, in the space at our disposal, to trace the career of all of them. Several came to an untimely end, but as the years went by each succeeding ship proved more efficient, and the first ship which was delivered to the Navy performed the notable flight of thirty-one hours.

To revert, for a moment, once more to the earlier ships--the fourth was wrecked and burned at Echterdingen in the same year in which she was completed. The fifth, which was the second military airship, was fitted with two 110 horse-power engines and also came to a tragic end, being destroyed by wind at Weilberg in 1910, and the following ship was burnt at Baden in the same year.

The seventh ship was the first passenger airship of the series, and was known as the Deutschland. By this time the capacity had increased to 536,000 cubic feet, and she was propelled by three 120 horse-power engines. She also fell a victim to the wind, and was wrecked in the Teutoberg Forest in 1910; and yet another was destroyed in the following year at Dusseldorf.

The tenth ship to be completed was the passenger ship Schwaben; her capacity was 636,500 cubic feet, and she had three 150 horse-power engines. This ship carried out her first flight in June, 1911, and was followed four months later by the Victoria Luise. The fourth passenger airship was known as the Hansa. These three ships were all in commission at the outbreak of war.

The first naval airship, L 1, mentioned above, was larger than any of these. The total length was 525 feet, diameter 50 feet, and cubic contents 776,000 cubic feet. Her hull framework in section formed a regular polygon of seventeen sides, and was built up of triangular aluminium girders. The gasbags were eighteen in number. This ship was fitted with three 170 horse-power Maybach engines, which were disposed as follows--one in the forward car, driving two two-bladed propellers; two in the after car, each driving a single four-bladed propeller. For steering purposes she had six vertical and eight horizontal planes. The total lift was 27 tons, with a disposable lift of 7 tons. Her speed was about 50 miles per hour, and she could carry fuel for about 48 hours. Her normal crew consisted of fourteen persons, including officers.

It will probably be remembered that the military Zeppelin Z III was compelled to make a forced landing in France. This ship was of similar construction to L 1, but of smaller volume, her capacity being 620,000 cubic feet. A trial flight was being carried out, and while above the clouds the crew lost their bearings. Descending they saw some French troops and rose again immediately. After flying for four hours they thought they must be safely over the frontier and, running short of petrol, made a landing--not knowing that they were still in France until too late. The airship was taken over by the French authorities.

Until the year 1916 the Zeppelin may be considered to have passed through three stages of design. Of the twenty-five ships constructed before the war, twenty-four were of the first type and one of the second. Each type possessed certain salient features, which, for simplicity, will be set out in the form of a tabulated statement, and may be useful for comparison when our own rigid airships are reviewed.

> Stage 1. Long parallel portion of hull with bluff nose and tail. External keel with walking way. Box rudders and elevators. Two cars. Four wing propellers.

Stage 2. Long parallel portion of hull with bluff nose, tail portion finer than in Stage I Internal keel walking way. Box rudders and elevators. Three cars, foremost for control only. Four wing propellers.

Stage 3. Shorter parallel portion of hull framework, bluff nose and tapering tail. Internal keel walking way. Balanced monoplane rudders and elevators. Three cars, foremost for control only. Two foremost cars close together and connected by a canvas joint to look like one car. Four engines and four propellers. One engine in forward car driving pusher propeller. Three engines in after car driving two wing and one pusher propeller.

To the second stage belongs naval airship L 2, which was destroyed by fire a month after completion in 1913. In 1916 a fourth stage made its appearance, of which the first ship was L 30, completed in May, and to which the ill-fated L 33 belonged. This type is known as the super-Zeppelin, and has been developed through various stage until L 70, the latest product before the armistice. In this stage the following are its main features:

Stage 4.
Short parallel portion of hull, long rounded bow and long tapering stern. In all respects a good streamline shape. Internal keel walking way. Balanced monoplane rudders and elevators. Five cars. Two forward (combined as in Stage 3), one aft, and two amidships abreast. Six engines and six propellers. The after one of the forecar and the sidecars each contain one engine driving direct a pusher propeller. The after car contains three engines, two of which drive two wing propellers; the third, placed aft, drives direct a pusher propeller. In this stage the type of girders was greatly altered.

A company known as the Schutte-Lanz Company was also responsible for the production of rigid airships. They introduced a design, which was a distinct departure from Zeppelin or anyone else. The hull framework was composed of wood, the girders being built up of wooden sections. The shape of these ships was much more of a true streamline than had been the Zeppelin practice, and it was on this model that the shape of the super-Zeppelin was based. These ships proved of use and took part in raids on this country, but the Company was taken over by the Government and the personnel was amalgamated with that engaged on Zeppelin construction during the war.

## ITALY

In 1908, Italy, stimulated by the progress made by other continental nations, commenced experimental work. Three types were considered for a commencement, the P type or Piccolo was the first effort, then followed the M type, which signifies "medium sized," and also the semirigid Forlanini.

In the Forlanini type the envelope is divided into several compartments with an internal rigid keel and to-day these ships are of considerable size, the most modern being over 600,000 cubic feet capacity. During the war, Italian airships were developed on entirely dissimilar lines to those in other countries. Both we and our Allies, and to a great extent the Germans, employed airships exclusively for naval operations; on the other hand, the Italian ships were utilized for bombing raids in conjunction with military evolutions.

For this reason height was of primary importance and speed was quite a secondary consideration, owing to the low velocity of prevailing winds in that country. Flights were never of long duration compared with those carried out by our airships. Height was always of the utmost importance, as the Italian ships were used for bombing enemy towns and must evade hostile gunfire. For this reason weight was saved in every possible manner, to increase the height of the "ceiling."

In addition to the types already mentioned, three other varieties have been constructed since the war--the Usuelli D.E. type and G class. The G class was a rigid design which has not been proceeded with, and, with this single exception, all are of a semirigid type in which an essentially non-rigid envelope is reinforced by a metal keel. In the Forlanini and Usuelli types the keel is completely rigid and assists in maintaining the shape of the envelopes, and in the Forlanini is enclosed within the envelope. In the other types the keel is in reality a chain of rigid links similar to that of a bicycle. The form of the envelope is maintained by the internal pressure and not by the keel, but the resistance of the latter to compression enables a lower pressure to be maintained than would be possible in a purely non-rigid ship.

The M type ship is of considerable size, the P smaller, while the D.E. is a small ship comparable to our own S.S. design. The review of these three countries brings the early history of airships to a conclusion. Little of importance was done elsewhere before the war, though Baldwin's airship is perhaps worthy of mention. It was built in America in 1908 by Charles Baldwin for the American Government. The capacity of the envelope was 20,000 cubic feet, she carried a crew of two, and her speed was 16 miles per hour. She carried out her trial flight in August, 1908, and was accepted by the American military authorities. During the war both the naval and military authorities became greatly interested in airships, and purchased several from the French and English. In addition to this a ship in design closely resembling the S.S. was built in America, but suffered from the same lack of experience which we did in the early days of airship construction.

We must now see what had been happening in this country in those fateful years before the bombshell of war exploded in our midst.

# CHAPTER III

## BRITISH AIRSHIPS BUILT BY PRIVATE FIRMS

It has been shown in the previous chapter that the development of the airship had been practically neglected in England prior to the twentieth century. Ballooning had been carried out both as a form of sport and also by the showman as a Saturday afternoon's sensational entertainment, with a parachute descent as the piece de resistance. The experiments in adapting the balloon into the dirigible had, however, been left to the pioneers on the Continent.

## PARTRIDGE'S AIRSHIP

It appears that in the nineteenth century only one airship was constructed in this country, which proved to be capable of ascending into the air and being propelled by its own machinery. This airship made its appearance in the year 1848, and was built to the designs of a man named Partridge. Very little information is available concerning this ship. The envelope was cylindrical in shape, tapering at each end, and was composed of a light rigid framework covered with fabric. The envelope itself was covered with a light wire net, from which the car was suspended. The envelope contained a single ballonet for regulating the pressure of the gas. Planes, which in design more nearly resembled sails, were used for steering purposes. In the car, at the after end, were fitted three propellers which were driven by compressed air.

Several trips of short duration were carried out in this airship, but steering was never successfully accomplished owing to difficulties encountered with the planes, and, except in weather of the calmest description, she may be said to have been practically uncontrollable.

## HUGH BELL'S AIRSHIP

In the same year, 1848, Bell's airship was constructed. The envelope of this ship was also cylindrical in shape, tapering at each end to a point, the length of which was 56 feet and the diameter 21 feet 4 inches. A keel composed of metal tubes was attached to the underside of the envelope from which the car was suspended. On either side of the car screw propellers were fitted to be worked by hand. A rudder was attached behind the car. It was arranged that trials should be carried out in the Vauxhall Gardens in London, but these proved fruitless.

## BARTON'S AIRSHIP

In the closing years of the nineteenth century appeared the forerunners of airships as they are to-day, and interest was aroused in this country by the performances of the ships designed by Santos-Dumont and Count Zeppelin. From now onwards we find various British firms turning their attention to the conquest of the air.

In 1903 Dr. Barton commenced the construction of a large non-rigid airship. The envelope was 176 feet long with a height of 43 feet and a capacity of 235,000 cubic feet; it was cylindrical in shape, tapering to a point at each end. Beneath the whole length of the cylindrical portion was suspended a bamboo framework which served as a car for the crew, and a housing for the motors supplying the motive power of the ship. This framework was suspended from the envelope by means of steel cables. Installed in the car were two 50 horse-power Buchet engines which were mounted at the forward and after ends of the framework. The propellers in themselves were of singular design, as they consisted of three pairs of blades mounted one behind the other. The were situated on each side of the car, two forward and two aft. The drive also include large friction clutches, and each engine was under separate control.

To enable the ship to be trimmed horizontally, water tanks were fitted at either end of the framework, the water being transferred from one to the other as was found necessary.

A series of planes was mounted at intervals along the framework to control the elevation of the ship.

This ship was completed in 1905 and was tried at the Alexandra Palace in the July of that year. She, unfortunately, did not come up to expectations, owing to the difficulty in controlling her, and during the trial flight she drifted away and was destroyed in landing.

## WILLOWS No. 1

From the year 1905 until the outbreak of war Messrs. Willows & Co. were engaged on the construction of airships of a small type, and considerable success attended their efforts. Each succeeding ship was an improvement on its predecessor, and flights were made which, in their day, created a considerable amount of interest.

In 1905 their first ship was completed. This was a very small non-rigid of only 12,500 cubic feet capacity. The envelope was made of Japanese silk, cylindrical in shape, with rather blunt conical ends. A long nacelle or framework, triangular in section and built up of light steel tubes, was suspended beneath the envelope by means of diagonally crossed suspensions.

A 7 horse-power Peugeot engine was fitted at the after end of the nacelle which drove a 10-feet diameter propeller. In front were a pair of swivelling tractor screws for steering the ship in the vertical and horizontal plane. No elevators or rudders were fixed to the ship.

## WILLOWS No. 2

The second ship was practically a semi-rigid. The envelope was over twice the capacity of the earlier ship, being of 29,000 cubic feet capacity. This envelope was attached to a keel of bamboo and steel, from which was suspended by steel cables a small car. At the after end of the keel was mounted a small rudder for the horizontal steering. For steering in the vertical plane two propellers were mounted on each side of the car, swivelling to give an upward or downward thrust. A 30 horse-power J.A.P. engine was fitted in this case. Several successful flights were carried out by this ship, of which the most noteworthy was from Cardiff to London.

## WILLOWS No. 3

No. 2, having been rebuilt and both enlarged and improved, became known as No. 3. The capacity of the envelope, which was composed of rubber and cotton, was increased to 32,000 cubic feet, and contained two ballonets. The gross lift amounted to about half a ton. As before, a 30 horse-power J.A.P. engine was installed, driving the swivelling propellers. These propellers were two-bladed with a diameter of 61 feet. The maximum speed was supposed to be 25 miles per hour, but it is questionable if this was ever attained.

This ship flew from London to Paris, and was the first British-built airship to fly across the Channel.

## WILLOWS No. 4

The fourth ship constructed by this firm was completed in 1912, and was slightly smaller than the two preceding ships. The capacity of the envelope in this instance was reduced to 24,000 cubic feet, but was a much better shape, having a diameter of 20 feet, which was gradually tapered towards the stern. A different material was also used, varnished silk being tried as an experiment. The envelope was attached to a keel on which was mounted the engine, a 35 horse-power Anzani, driving two swivelling four-bladed propellers. From the keel was suspended a torpedo-shaped boat car in which a crew of two was accommodated. Originally a vertical fin and rudder were mounted at the stern end of the keel, but these were later replaced by fins on the stern of the envelope.

This ship was purchased by the naval authorities, and after purchase was more or less reconstructed, but carried out little flying. At the outbreak of war she was lying deflated in the shed at Farnborough. As will be seen later, this was the envelope which was rigged to the original experimental S.S. airship in the early days of 1915, and is for this reason, if for no other, particularly interesting.

## WILLOWS No. 5

This ship was of similar design, but of greater capacity. The envelope, which was composed of rubber-proofed fabric, gave a volume of 50,000 cubic feet, and contained two ballonets. A 60 horsepower engine drove two swivelling propellers at an estimated speed of 38 miles per hour. She was constructed at Hendon, from where she made several short trips.

## MARSHALL FOX'S AIRSHIP

In the early days of the war an airship was constructed by Mr. Marshall Fox which is worthy of mention, although it never flew. It was claimed that this ship was a rigid airship, although from its construction it could only be looked upon as a non-rigid ship, having a wooden net-work around its envelope. The hull was composed of wooden transverse frames forming a polygon of sixteen sides, with radial wiring fitted to each transverse frame. The longitudinal members were spiral in form and were built up of three-ply lathes. A keel of similar construction ran along the under side of the hull which carried the control position and compartments for two Green engines, one of 40 horse-power, the other of 80 horse-power, together with the petrol, bombs, etc.

In the hull were fitted fourteen gasbags giving a total capacity of 100,000 cubic feet. The propeller drive was obtained by means of a wire rope. The gross lift of the ship was 4,276 lb., and the weight of the structure, complete with engines, exceeded this.

It became apparent that the ship could never fly, and work was suspended. She was afterwards used for carrying out certain experiments and at a later date was broken up.

Apart from the various airships built under contract for the Government there do not appear to be any other ships built by private firms which were completed and actually flew. It is impossible to view this lack of enterprise with any other feelings than those of regret, and it was entirely due to this want of foresight that Great Britain entered upon the World War worse equipped, as regards airships, than the Central Empires or any of the greater Allied Powers.

# CHAPTER IV

## BRITISH ARMY AIRSHIPS

The French and German military authorities began to consider airships as an arm of the Service in the closing years of the nineteenth century, and devoted both time and considerable sums of money in the attempt to bring them to perfection. Their appearance in the British Army was delayed for many years on account of the expense that would be incurred in carrying out experiments. In 1902, Colonel Templer, at that time head of the Balloon Section, obtained the necessary sanction to commence experiments, and two envelopes of gold-beaters skin of 50,000 cubic feet capacity were built. With their completion the funds were exhausted, and nothing further done until 1907.

## NULLI SECUNDUS I

In 1907 the first complete military airship in England was built, which bore the grandiloquent title of Nulli Secundus. One of the envelopes constructed by Colonel Templer was used: it was cylindrical in shape with spherical ends. Suspended beneath the envelope by means of a net and four broad silk bands was a triangular steel framework or keel from which was slung a small car. A 50 horsepower Antoinette engine was situated in the forward part of the car which drove two metal-bladed propellers by belts. At the after part of the keel were fitted a rudder and small elevators, and two pairs of movable horizontal planes were also fitted forward. It is remarkable that no stabilizing surfaces whatsoever were mounted. The envelope was so exceedingly strong that a high pressure of gas could be sustained, and ballonets were considered unnecessary, but relief valves were employed. The first flight took place in September and was fairly successful. Several were made afterwards, and in October she was flown over London and landed at the Crystal Palace. The flight lasted 3 hours and 25 minutes, which constituted at the time a world's record. Three days later, owing to heavy winds, the ship had to be deflated and was taken back to Farnborough.

## NULLI SECUNDUS II

In 1908 the old ship was rebuilt with several modifications. The envelope was increased in length and was united to the keel by means of a covering of silk fabric in place of the net, four suspension bands being again used. A large bow elevator was mounted which made the ship rather unstable. A few flights were accomplished, but the ship proved of little value and was broken up.

## BABY

This little airship made its first appearance in the spring of 1909. The envelope was fish-shaped and composed of gold-beater's skin, with a volume of 21,000 cubic feet. One ballonet was contained in the envelope which, at first, had three inflated fins to act as stabilizers. These proved unsatisfactory as they lacked rigidity, and were replaced after the first inflation by the ordinary type. Two 8 horsepower 3-cylinder Berliet engines were mounted in a long car driving a simple propeller, and at a later date were substituted by a R.E.P. engine which proved most unsatisfactory. During the autumn permission was obtained to enlarge the envelope and fit a more powerful engine.

## BETA

Beta was completed in May, 1910. The envelope was that of the Baby enlarged, and now had a volume of 35,000 cubic feet. The car was composed of a long frame, having a centre compartment for the crew and engines, which was the standard practice at that time for ships designed by the Astra Company. A 35 horse-power Green engine drove two wooden two-bladed propellers by chains. The ship was fitted with an unbalanced rudder, while the elevators were in the front of the frame. This ship was successful, and in June flew to London and back, and in September took part in the Army manoeuvres, on one occasion being in the air for 7 3/4 hours without landing, carrying a crew of three. Trouble was experienced in the steering, the elevators being situated too near the centre of the ship to be really efficient and were altogether too small.

In 1912, Beta, having been employed regularly during the previous year, was provided with a new car having a Clerget engine of 45 horse-power. In 1913 she was inflated for over three months and made innumerable flights, on one occasion carrying H.R.H. the Prince of Wales as passenger. She had at that time a maximum speed of 35 miles per hour, and could carry fuel for about eight hours with a crew of three.

## GAMMA

In 1910 the Gamma was also completed. This was a much bigger ship with an envelope of 75,000 cubic feet capacity, which, though designed in England, had been built by the Astra Company in Paris. The car, as in Beta, was carried in a long framework suspended from the envelope. This portion of the ship was manufactured in England, together with the machinery. This consisted of an 80 horse-power Green engine driving swivelling propellers, the gears and shafts of which were made by Rolls Royce. The engine drove the propeller shafts direct, one from each end of the crankshaft.

Originally the envelope was fitted with inflated streamline stabilizers on either side, but at a later date these were replaced by fixed stabilizing planes. At the same time the Green engine was removed and two Iris engines of 45 horse-power were installed, each driving a single propeller. There were two pairs of elevators, each situated in the framework, one forward, the other aft. In 1912, having been rigged to a new envelope of 101,000 cubic feet capacity, the ship took part in the autumn manoeuvres, and considerable use was made of wireless telegraphy.

In a height reconnaissance the pilot lost his way, and running out of petrol drifted all night, but was safely landed. When returning to Farnborough the rudder controls were broken and the ship was ripped. In this operation the framework was considerably damaged. When repairs were being carried out the elevators were removed from the car framework and attached to the stabilizing fins in accordance with the method in use to-day.

## CLEMENT-BAYARD

In 1910 it was arranged by a committee of Members of Parliament that the Clement-Bayard firm should send over to England a large airship on approval, with a view to its ultimate purchase by the War Office, and a shed was erected at Wormwood Scrubs for its accommodation. This ship arrived safely in October, but was very slow and difficult to control. The envelope, moreover, was of exceedingly poor quality and consumed so much gas that it was decided to deflate it. She was taken to pieces and never rebuilt.

## LEBAUDY

About the same time, interest having been aroused in this country by the success of airships on the Continent, the readers of the Morning Post subscribed a large sum to purchase an airship for presentation to the Government. This was a large ship of 350,000 cubic feet capacity and was of semi-rigid design, a long framework being suspended from the envelope which supported the weight of the car. It had two engines of 150 horse-power which developed a speed of about 32 miles per hour. The War Office built a shed at Farnborough to house it, and in accordance with dimensions given by the firm a clearance of 10 feet was allowed between the top of the ship and the roof of the shed. Inconceivable as it may sound, the overall height of the ship was increased by practically 10 feet without the War Office being informed. The ship flew over and was landed safely, but on being taken into the shed the envelope caught on the roof girders, owing to lack of headroom, and was ripped from end to end. The Government agreed to increase the height of the shed and the firm to rebuild the ship. This was completed in March, 1911, and the ship was inflated again. On carrying out a trial flight, having made several circuits at 600 feet, she attempted to land, but collided with a house and was completely wrecked. This was the end of a most unfortunate ship, and her loss was not regretted.

## DELTA

Towards the end Of 1910 the design was commenced of the ship to be known as the Delta, and in 1911 the work was put in hand. The first envelope was made of waterproofed silk. This proved a failure, as whenever the envelope was put up to pressure it invariably burst. Experiments were continued, but no good resulting, the idea was abandoned and a rubber-proofed fabric envelope was constructed of 173,000 cubic feet volume. This ship was inflated in 1912. The first idea was to make the ship a semi-rigid by lacing two flat girders to the sides of the envelope to take the weight of the car. This idea had to be abandoned, as in practice, when the weight of the car was applied, the girders buckled. The ship was then rigged as a non-rigid. A novelty was introduced by attaching a rudder flap to the top stabilizing fin, but as it worked somewhat stiffly it was later on removed. This ship took part in the manoeuvres of 1912 and carried out several flights. She proved to be exceedingly fast, being capable of a speed of 44 miles per hour. In 1913 she was completely re-rigged and exhibited at the Aero Show, but the re-designed rigging revealed various faults and it was not until late in the year that she carried out her flight trials. Two rather interesting experiments were made during these flights. In one a parachute descent was successfully accomplished; and in another the equivalent weight of a man was picked up from the ground without assistance or landing the ship.

## ETA

The Eta was somewhat smaller than the Delta, containing only 118,000 cubic feet of hydrogen, and was first inflated in 1913. The envelope was composed of rubber-proofed fabric and a long tapering car was suspended, this being in the nature of a compromise between the short car of the, Delta and the long framework gear of the Gamma. Her engines were two 80 horse-power Canton-Unne, each driving one propeller by a chain. This ship proved to be a good design and completed an eight-hour trial flight in September. On her fourth trial she succeeded in towing the disabled naval airship No. 2 a distance of fifteen miles. Her speed was 42 miles per hour, and she could carry a crew of five with fuel for ten hours.

On January 1st, 1914, the Army disbanded their Airship Section, and the airships Beta, Gamma, Delta and Eta were handed over to the Navy together with a number of officers and men.

# CHAPTER V

## EARLY DAYS OF THE NAVAL AIRSHIP SECTION--PARSEVAL AIRSHIPS, ASTRA-TORRES TYPE, ETC.

The rapid development of the rigid airships in Germany began to create a considerable amount of interest in official circles. It was realized that those large airships in the future would be invaluable to a fleet for scouting purposes. It was manifest that our fleet, in the event of war, would be gravely handicapped by the absence of such aerial scouts, and that Germany would hold an enormous advantage if her fleet went to sea preceded by a squadron of Zeppelin airships.

The Imperial Committee, therefore, decided that the development of the rigid airship should be allotted to the Navy, and a design for Rigid Airship No. 1 was prepared by Messrs. Vickers in conjunction with certain naval officers in the early part of 1909.

As will be seen later this ship was completed in 1911, but broke in two in September of that year and nothing more was done with her. In February, 1912, the construction of rigid airships was discontinued, and in March the Naval Airship section was disbanded.

In September, 1912, the Naval Airship section was once more reconstituted and was stationed at Farnborough. The first requirements were airships, and owing to the fact that airship construction was so behindhand in this country, in comparison with the Continent, it was determined that purchases should be made abroad until sufficient experience had been gained by British firms to enable them to compete with any chance of success against foreign rivals.

First a small non-rigid, built by Messrs. Willows, was bought by the Navy to be used for the training of airship pilots. In addition an Astra-Torres airship was ordered from France. This was a ship of 229,450 cubic feet capacity and was driven by twin Chenu engines of 210 horse-power each. She carried a crew of six, and was equipped with wireless and machine guns. The car could be moved fore and aft for trimming purposes, either by power or by hand. This was, however, not satisfactory, and was abandoned.

In April 1918, Messrs. Vickers were asked to forward proposals for a rigid airship which afterwards became e known as No. 9. Full details of the vicissitudes connected with this ship will be given in the chapter devoted to Rigid Airships.

In July, approval was granted for the construction of six non-rigid ships. Three of these were to be of the German design of Major von Parseval and three of the Forlanini type, which was a semi-rigid design manufactured in Italy. The order for the Parsevals was placed with Messrs. Vickers and for the Forlaninis with Messrs. Armstrong.

The Parseval airship was delivered to this country and became known as No. 4; a second ship of the same type was also building when war broke out; needless to say this ship was never delivered. At a later date Messrs. Vickers, who had obtained the patent rights of the Parseval envelope, completed the other two ships of the order.

The Forlanini ship was completing in Italy on the declaration of war and was taken over by the Italians; Messrs. Armstrong had not commenced work on the other two. These ships, although allocated numbers, never actually came into being.

## PARSEVAL AIRSHIP No. 4

This airship deserves special consideration for two reasons; firstly, on account of the active-service flying carried out by it during the first three years of the war, and, secondly, for its great value in training of the officers and men who later on became the captains and crews of rigid airships.

The Parseval envelope is of streamline shape which tapers to a point at the tail, and in this ship was of 300,000 cubic feet capacity. The system of rigging being patented, can only be described in very general terms. The suspensions carrying the car are attached to a large elliptical rigging band which is formed under the central portion of the envelope. To this rigging band are attached the trajectory bands which pass up the sides and over the top of the envelope, sloping away from the centre at the bottom towards the nose and tail at the top. The object of this is to distribute the load fore and aft over the envelope. These bands, particularly at the after end of the ship, follow a curved path, so that they become more nearly vertical as they approach the upper surface of the envelope. This has the effect of bringing the vertical load on the top of the envelope; but a greater portion of the compressive force comes on the lower half, where it helps to resist the bending moment due to the unusually short suspensions. A single rudder plane and the ordinary elevator planes were fitted to the envelope. A roomy open car was provided for this ship, composed of a duralumin framework and covered with duralumin sheeting. Two 170 horse-power Maybach engines were mounted at the after end of the car, which drove two metal-bladed reversible propellers. These propellers were later replaced by standard four-bladed wooden ones and a notable increase of speed was obtained.

Two officers and a crew of seven men were carried, together with a wireless installation and armament.

This airship, together with No. 3, took part in the great naval review at Spithead, shortly before the commencement of the war, and in addition to the duties performed by her in the autumn of 1914, which are mentioned later, carried out long hours of patrol duty from an east coast station in the summer of 1917. In all respects she must be accounted a most valuable purchase.

## PARSEVAL AIRSHIPS 5, 6 and 7

Parseval No. 5 was not delivered by Germany owing to the war, so three envelopes and two cars were built by Messrs. Vickers on the design of the original ship. These were delivered somewhat late in the war, and on account of the production of the North Sea airship with its greater speed were not persevered with. The dimensions of the envelopes were somewhat increased, giving a cubic capacity of 325,000 cubic feet. Twin Maybach engines driving swivelling propellers were installed in the car, which was completely covered in, but these ships were slow in comparison with later designs, and were only used for the instruction of officers and men destined for the crews of rigid airships then building.

An experimental ship was made in 1917 which was known as Parseval 5; a car of a modified coastal pattern with two 240 horse-power Renault engines was rigged to one of envelopes. During a speed trial, this ship was calculated to have a ground speed of 50 to 53 miles per hour. The envelope, however, consumed an enormous amount gas and for this reason the ship was deflated and struck off the list of active ships.

This digression on Parseval airships has anticipated events somewhat, and a return must now be made to earlier days.

Two more Astra-Torres were ordered from France, one known as No. 8, being a large ship of 4,00,000 cubic feet capacity. She was fitted with two Chenu engines of 240 horse-power, driving swivelling propellers. This ship was delivered towards the end of the year 1914. The second Astra was of smaller capacity and was delivered, but as will be seen later, was never rigged, the envelope being used for the original coastal ship and the car slung to the envelope of the ex-army airship Eta.

On January 1st, 1914, an important event took place: the Army disbanded their airship service, and the military ships together with certain officers and men were transferred to the Naval Air Service.

Before proceeding further, it may be helpful to explain the system by which the naval airships have been given numbers. These craft are always known by the numbers which they bear, and the public is completely mystified as to their significance whenever they fly over London or any large town. It must be admitted that the method is extremely confusing, but the table which follows should help to elucidate the matter. The original intention was to designate each airship owned by the Navy by a successive number. The original airship, the rigid Mayfly, was known as No. 1, the Willows airship No. 2, and so on. These numbers were allocated regardless of type and as each airship was ordered, consequently some of these ships, for example the Forlaninis, never existed. That did not matter, however, and these numbers were not utilized for ships which actually were commissioned. On the transfer of the army airships, four of these, the Beta, Gamma, Delta and Eta, were given their numbers as they were taken over, together with two ships of the Epsilon class which were ordered from Messrs. Rolls Royce, but never completed. In this way it will be seen that numbers 1 to 22 are accounted for.

In 1915 it was decided to build a large number of small ships for anti-submarine patrol, which were called S.S.'s or Submarine Scouts. It was felt that it would only make confusion worse confounded if these ships bore the original system of successive numbering and were mixed up with those of later classes which it was known would be produced as soon as the designs were completed. Each of these ships was accordingly numbered in its own class, S.S., S.S.P., S.S. Zero, Coastal, C Star and North Sea, from 1 onwards as they were completed.

In the case of the rigids, however, for some occult reason the old system of numbering was persisted in. The letter R is prefixed before the number to show that the ship is a rigid. Hence we have No. 1 a rigid, the second rigid constructed is No. 9, or R 9, and the third becomes R 23. From this number onwards all are rigids and are numbered in sequence as they are ordered, with the exception of the last on the list, which is a ship in a class of itself. This ship the authorities, in their wisdom, have called R 80--why, nobody knows.

With this somewhat lengthy and tedious explanation the following table may be understood:

> No. Type. Remarks. 1. Rigid Wrecked, Sept. 24, 1911. 2. Willows Became S.S. 1. 3. Astra-Torres Deleted, May 1916. 4. Parseval Deleted, July, 1917. 5. Parseval Never delivered from Germany. (Substitute ship built by Messrs. Vickers). 6. Parseval Built by Messrs. Vickers. 7. Parseval Built by Messrs. Vickers. 8. Astra-Torres Deleted, May, 1916. 9. Rigid Deleted, June, 1918. 10. Astra-Torres Envelope used for C 1. 11. Forlanini Never delivered owing to war. 12. Forlanini Never delivered owing to war. 13. Forlanini Never delivered owing to war. 14. Rigid Never built. 15. Rigid Never built. 16. Astra-Torres See No. 8. 17. Beta Transferred from Army. Deleted, May, 1916. 18. Gamma Deleted, May, 1916. 19. Delta Deleted, May, 1916. 20. Eta Transferred from the Army. Fitted with car from No. 10. Deleted May, 1916. 21. Epsilon Construction cancelled May, 1916. 22. Epsilon Construction cancelled May,

1916. 23. Rigid 23 Class. 24. Rigid 23 Class. 25. Rigid 23 Class. 26. Rigid 23 Class. 27. Rigid 23x Class. 28. Rigid 23x Class. Never completed. 29. Rigid 23x Class. 30. Rigid 23x Class. Never completed. 31. Rigid 31 Class. 32. Rigid 31 Class, building. 33. Rigid 33 Class. 34. Rigid 33 Class. 35. Rigid Cancelled. 36. Rigid Building. 37. Rigid Building. 38. Rigid Building. 39. Rigid Building. 40. Rigid Building. 80. Rigid Building.

In August, 1914, Europe, which had been in a state of diplomatic tension for several years, was plunged into the world war. The naval airship service at the time was in possession of two stations, Farnborough and Kingsnorth, the latter in a half-finished condition. Seven airships were possessed, Nos. 2, 3 and 4, and the four ex-army ships--Beta, Gamma, Delta and Eta--and of these only three, Nos. 3, 4 and the Beta, were in any condition for flying. Notwithstanding this, the utmost use was made of the ships which were available.

On the very first night of the war, Nos. 3 and 4 carried out a reconnaissance flight over the southern portion of the North Sea, and No. 4 came under the fire of territorial detachments at the mouth of the Thames on her return to her station. These zealous soldiers imagined that she was a German ship bent on observation of the dockyard at Chatham.

No. 3 and No. 4 rendered most noteworthy service in escorting the original Expeditionary Force across the Channel, and in addition to this No. 4 carried out long patrols over the channel throughout the following winter.

No. 17 (Beta) also saw active service, as she was based for a short period early in 1915 at Dunkirk, and was employed in spotting duties with the Belgian artillery near Ostend.

The Gamma and the Delta were both lying deflated at Farnborough at the outbreak of the war, and in the case of the latter the car was found to be beyond repair, and she was accordingly deleted. The Gamma was inflated in January, 1915, and was used for mooring experiments.

The Eta, having been inflated and deflated several times owing to the poor quality of the envelope, attempted to fly to Dunkirk in November, 1914. She encountered a snowstorm near Redhill and was compelled to make a forced landing. In doing this she was so badly damaged as to be incapable of repair, and at a later date was deleted.

No. 8, which was delivered towards the end of 1914, was also moored out in the open for a short time near Dunkirk, and carried out patrol in the war zone of the Belgian coast.

So ends the story of the Naval Airship Service before the war.

With the submarine campaign ruthlessly waged by the Germans from the spring of 1915 and onwards, came the airship's opportunity, and the authorities grasped the fact that, with development, here was the weapon to defeat the most dangerous enemy of the Empire. The method of development and the success attending it the following chapters will show.

# CHAPTER VI

### NAVAL AIRSHIPS.--THE NON-RIGIDS--S.S. TYPE

The development of the British airships of to-day may be said to date from February 28th, 1915. On that day approval was given for the construction of the original S.S. airship.

At this time the Germans had embarked upon their submarine campaign, realizing, with the failure of their great assaults on the British troops in Flanders, that their main hope of victory lay in starving Great Britain into surrender. There is no doubt that the wholesale sinking of our merchant shipping was sufficient to cause grave alarm, and the authorities were much concerned to devise means of minimizing, even if they could not completely eliminate the danger. One proposal which was adopted, and which chiefly concerns the interests of this book, was the establishment of airship stations round the coasts of Great Britain. These stations were to be equipped with airships capable of patrolling the main shipping routes, whose functions were to search for submarines and mines and to escort shipping through the danger zones in conjunction with surface craft.

Airship construction in this country at the time was, practically speaking, non-existent. There was no time to be wasted in carrying out long and expensive experiments, for the demand for airships which could fulfil these requirements was terribly urgent, and speed of construction was of primary importance. The non-rigid design having been selected for simplicity in construction, the expedient was tried of slinging the fuselage of an ordinary B.E. 2C aeroplane, minus the wings, rudder and elevators and one or two other minor fittings, beneath an envelope with tangential suspensions, as considerable experience had been gained already in a design of this type.

For this purpose the envelope of airship No. 2, which was lying deflated in the shed at Farnborough, was rushed post haste to Kingsnorth, inflated and rigged to the fuselage prepared for it. The work was completed with such despatch that the airship carried out her trial flight in less than a fortnight from approval being granted to the scheme. The trials were in every way most satisfactory, and a large number of ships of this design was ordered immediately. At the same time two private firms were invited to submit designs of their own to fulfil the Admiralty requirements. One firm's design, S.S. 2, did not fulfil the conditions laid down and was put out of commission; the other, designed by Messrs. Armstrong, was sufficiently successful for them to receive further orders. In addition to these a car was designed by Messrs. Airships Ltd., which somewhat resembled a Maurice Farman aeroplane body, and as it appeared to be suitable for the purpose, a certain number of these was also ordered.

About this period the station at Farnborough was abandoned by the Naval Airship Service to make room for the expansion of the military aeroplane squadrons. The personnel and airships were transferred to Kingsnorth, which became the airship headquarters.

The greatest energy was displayed in preparing the new stations, which were selected as bases for the airships building for this anti-submarine patrol. Small sheds, composed of wood, were erected with almost incredible rapidity, additional personnel was recruited, stores were collected, huts built for their accommodation and that of the men, and by the end of the summer the organization was so complete that operations were enabled to commence.

The S.S., or submarine scout, airship proved itself a great success. Beginning originally with a small programme the type passed through various developments until, at the conclusion of the war, no fewer than 150 ships of various kinds had been constructed. The alterations which took place and the improvements effected thereby will be considered at some length in the following pages.

## S.S.B.E. 2C

The envelope of the experimental ship S.S. 1 was only of 20,500 cubic feet capacity; for the active-service ships, envelopes of similar shape of 60,000 cubic feet capacity were built. The shape was streamline, that is to say, somewhat blunt at the nose and tapering towards the tail, the total length being 143 feet 6 inches, with a maximum diameter of 27 feet 9 inches.

The gross lift of these ships with 98% pure gas at a temperature of 60 degrees Fahrenheit and barometer 30 inches, is 4,180 lb. The net lift available for crew, fuel, ballast, armament, etc., 1,434 lb., and the disposable lift still remaining with crew of two on board and full tanks, 659 lb.

The theoretical endurance at full speed as regards petrol consumption is a little over 8 hours, but in practice it is probable that the oil would run short before this time had been reached. At cruising speed, running the engine at 1,250 revolutions, the consumption is at the rate of 3.6 gallons per hour, which corresponds to an endurance of 16 1/2 hours.

With the engine running at 1,800 revolutions, a speed of 50.6 miles per hour has been reached by one of these ships, but actually very few attained a greater speed than 40 miles per hour.

The envelopes of S.S. airships are composed of rubber-proofed fabric, two fabrics being used with rubber interposed between and also on the inner or gas surface. To render them completely gastight and as impervious to the action of the weather, sun, etc., as possible, five coats of dope are applied externally, two coats of Delta dope, two of aluminium dope and one of aluminium varnish applied in that order.

One ripping panel is fitted, which is situated on the top of the envelope towards the nose. It has a length of 14 feet 5 inches and a breadth of about 8 inches. The actual fabric which has to be torn away overlaps the edge of the opening on each side. This overlap is sewn and taped on to the envelope and forms a seam as strong and gastight as any other portion of the envelope. Stuck on this fabric is a length of biased fabric 8 1/4 inches wide. These two strips overlap the opening at the forward end by about three feet. At this end the two strips are loose and have a toggle inserted at the end to which the ripping cord is tied. The ripping cord is operated from the car. It is led aft from the ripping panel to a pulley fixed centrally over the centre of the car, from the pulley the cord passes round the side of the envelope and through a gland immediately below the pulley.

The nose of the envelope is stiffened to prevent it blowing in. For this purpose 24 canes are fitted in fabric pockets around the nose and meet at a point 2 1/4 inches in front of the nose. An aluminium conical cap is fitted over the canes and a fabric nose cap over the whole.

Two ballonets are provided, one forward and one aft, the capacity of each being 6,375 cubic feet. The supply of air for filling these is taken from the propeller draught by a slanting aluminium tube to the underside of the envelope, where it meets a longitudinal fabric hose which connects the two ballonet air inlets. Non-return fabric valves known as crab-pots are fitted in this fabric hose on either side of their junction with the air scoop. Two automatic air valves are fitted to the underside of the envelope, one for each ballonet. The air pressure tends to open the valve instead of keeping it shut and to counteract this the spring of the valve is inside the envelope. The springs are set to open at a pressure of 25 to 28 mm.

Two gas valves are also fitted, one on the top of the envelope, the other at the bottom. The bottom gas valve spring is set to open at 30 to 35 mm. pressure, the top valve is hand controlled only.

These valves are all very similar in design. They consist of two wooden rings, between which the envelope is gripped, and which are secured to each other by studs and butterfly nuts. The valve disc, or moving portion of the valve, is made of aluminium and takes a seating on a thin india rubber ring stretched between a metal rod bent into a circle of smaller diameter than the valve opening and the wooden ring of the valve. When it passes over the wooden ring it is in contact with the envelope fabric and makes the junction gastight. The disc is held against the rubber by a compressed spring.

The valve cords are led to the pilot's seat through eyes attached to the envelope.

The system of rigging or car suspension is simplicity itself and is tangential to the envelope. On either side there are six main suspensions of 25 cwt. stranded steel cable known as "C" suspensions. Each "C" cable branches into two halves known as the "B" bridles, which in turn are supported at each end by the bridles known as "A." The ends of the "A" bridles are attached to the envelope by means of Eta patches. These consist of a metal D-shaped fitting round which the rigging is spliced and through which a number of webbing bands are passed which are spread out fanwise and solutioned to the envelope. It will thus be seen that the total load on each main suspension is proportionally taken up by each of the four "A" bridles, and that the whole weight of the car is equally distributed over the greater part of the length of the envelope. Four handling guys for manoeuvering the ship on the ground are provided under the bow and under the stem. A group of four Eta patches are placed close together, which form the point of attachment for two guys in each case. The forward of these groups of Eta patches forms the anchoring point. The bridle, consisting of 25 cwt. steel cable, is attached here and connected to the forepart of the skids of the car. The junction of this bridle with the two cables from the skids forms the mooring point and there the main trail rope is attached. This is 120 feet long and composed of 2-inch manilla. This is attached, properly coiled, to the side of the car and is dropped by a release gear. It is so designed that when the airship is held in a wind by the trail rope the strain is evenly divided between the envelope and the car. The grapnel carried is fitted to a short length of rope. The other end of the rope has an eye, and is fitted to slide down the main trail rope and catch on a knot at the end.

For steering and stabilizing purposes the S.S. airship was originally designed with four fins and rudders, which were to be set exactly radial to the envelope. In some cases the two lower fins and rudders were abandoned, and a single vertical fin and rudder fitted centrally under the envelope were substituted. The three planes are identical in size and measure 16 feet by 8 feet 6 inches, having a gross stabilizing area of 402 1/2 square feet.

They are composed of spruce and aluminium and steel tubing braced with wire and covered by linen doped and varnished when in position.

The original rudders measured 3 feet by 8 feet 6 inches. In the case, however, of the single plane being fitted, 4-feet rudders are invariably employed. Two kingposts of steel tube are fitted to each plane and braced with wire to stiffen the whole structure.

The planes are attached to the envelope by means of skids and stay wires. The skids, composed of spruce, are fastened to the envelope by eight lacing patches.

The car, it will be remembered, is a B.E. 2C fuselage stripped of its wings, rudders and elevators, with certain other fittings added to render it suitable for airship work. The undercarriage is formed of two ash skids, each supported by three struts. The aeroplane landing wheels, axle and suspensions are abandoned.

In the forward end of the fuselage was installed a 75 horse-power air cooled Renault engine driving a single four-bladed tractor propeller through a reduction gear of 2 to 1. The engine is of the 8-cylinder V type, weighing 438 lb. with a bore of 96 mm. and a stroke of 120 mm. The Claudel-Hobson type of carburettor is employed with this engine. The type of magneto used is the Bosch D.V.4, there being one magneto for each line of cylinders. In the older French Renaults the Bosch H.L.8 is used, one magneto supplying the current to all the plugs. Petrol is carried in three tanks, a gravity and intermediate tank as fitted to the original aeroplane, and a bottom tank placed underneath the front seat of the car. The petrol is forced by air pressure from the two lower tanks into the gravity tank and is obtained by a hand pump fitted outside the car alongside the pilot's seat. The oil tank is fitted inside the car in front of the observer.

The observer's seat is fitted abaft the engine and the pilot's seat is aft of the observer. The observer, who is also the wireless operator, has the wireless apparatus fitted about his seat. This consists of a receiver and transmitter fitted inside the car, which derives power from accumulator batteries. The aerial reel is fitted outside the car. During patrols signals can be sent and received up to and between 50 and 60 miles.

The pilot is responsible for the steering and the running of the engine, and the controls utilized are the fittings supplied with the aeroplane. Steering is operated by the feet and elevating by a vertical wheel mounted in a fore and aft direction across the seat. The control wires are led aft inside the fairing of the fuselage to the extreme end, whence they pass to the elevators and rudders.

The instrument board is mounted in front of the pilot. The instruments comprise a watch, an air-speed indicator graduated in knots, an aneroid reading to 10,000 feet, an Elliott revolution counter, a Clift inclinometer reading up to 20 degrees depression or elevation, a map case with celluloid front.

There are in addition an oil pressure gauge, a petrol pressure gauge, a glass petrol level and two concentric glass pressure gauges for gas pressure.

The steering compass is mounted on a small wooden pedestal on the floor between the pilot's legs.

The water-ballast tank is situated immediately behind the pilot's seat and contains 14 gallons of water weighing 140 lbs. The armament consists of a Lewis gun and bombs. The bombs are carried in frames suspended about the centre of the undercarriage. The bomb sight is fitted near the bomb releasing gear outside the car on the starboard side adjacent to the pilot's seat. The Lewis gun, although not always carried on the early S.S. airships, was mounted on a post alongside the pilot's seat.

## S.S. MAURICE FARMAN

For this type of S.S. the cars were built by Messrs. Airships Ltd. In general appearance they resemble the Maurice Farman aeroplane and were of the pusher type; 60,000 and in later cases 70,000 cubic feet envelopes were rigged to these ships, which proved to be slightly slower than the B.E. 2C type, but this was compensated for owing to the increased comfort provided for the crew, the cars being more roomy and suitable for airship work in every way.

The system of rigging to all intents and purposes is the same in all types of S.S. ships, the suspensions being adjusted to suit the different makes of car.

In these ships the pilot sits in front, and behind him is the wireless telegraphy operator; in several cases a third seat was fitted to accommodate a passenger or engineer; dual rudder and elevator controls are provided for the pilot and observer.

The engine is mounted aft, driving a four-bladed pusher propeller, with the petrol tanks situated in front feeding the carburettors by gravity. The engines used are Rolls Royce Renaults, although in one instance a 75 horse-power Rolls Royce Hawk engine was fitted, which assisted in making an exceedingly useful ship.

## S.S. ARMSTRONG WHITWORTH

The car designed by Messrs. Armstrong Whitworth is of the tractor type and is in all ways generally similar to the B.E. 2C. The single-skid landing chassis with buffers is the outstanding difference. These cars had to be rigged to 70,000 cubic feet envelopes otherwise the margin of lift was decidedly small. A water-cooled 100 horsepower Green engine propelled the ship, and a new feature was the disposition of petrol, which was carried in two aluminium tanks slung from the envelope and fed through flexible pipes to a two-way cock and thence to the carburettors. These tanks, which were supported in a fabric sling, showed a saving in weight of 100 lb. compared with those fitted in the B.E. 2C.

For over two years these three types of S.S. ships performed a great part of our airship patrol and gave most excellent results.

Owing to the constant patrol which was maintained whenever weather conditions were suitable, the hostile submarine hardly dared to show her periscope in the waters which were under observation. In addition to this, practically the whole of the airship personnel now filling the higher positions, such as Captains of Rigids and North Seas, graduated as pilots in this type of airship. From these they passed to the Coastal and onwards to the larger vessels.

As far as is known the height record for a British airship is still held by an S.S.B.E. 2C, one of these ships reaching the altitude of 10,300 feet in the summer of 1916.

The Maurice Farman previously mentioned as being fitted with the Hawk engine, carried out a patrol one day of 18 hours 20 minutes. In the summer of 1916 one of the Armstrong ships was rigged to an envelope doped black and sent over to France. While there she carried out certain operations at night which were attended with success, proving that under certain circumstances the airship can be of value in operating with the military forces over land.

## S.S.P.

In 1916 the design was commenced for an S.S. ship which should have a more comfortable car and be not merely an adaptation of an aeroplane body. These cars, which were of rectangular shape with a blunt nose, were fitted with a single landing skid aft, and contained seats for three persons.

The engine, a 100 horse-power water-cooled Green, was mounted on bearers aft and drove a four-bladed pusher propeller. The petrol was carried in aluminium tanks attached by fabric slings to the axis of the envelope.

Six of these ships were completed in the spring of 1917 and were quite satisfactory, but owing to the success achieved by the experimental S.S. Zero it was decided to make this the standard type of S.S. ship, and with the completion of the sixth the programme of the S.S.P's was brought to a close.

These ships enjoyed more than, perhaps, was a fair share of misfortune, one was wrecked on proceeding to its patrol station and was found to be beyond repair, and another was lost in a snowstorm in the far north. The remainder, fitted at a later date with 75 horse-power Rolls Royce engines, proved to be a most valuable asset to our fleet of small airships.

## S.S. ZERO

The original S.S. Zero was built at a south-coast station by Air Service labour, and to the design of three officers stationed there. The design of the car shows a radical departure from anything that had been previously attempted, and as a model an ordinary boat was taken. In shape it is as nearly streamline as is practicable, having a keel and ribs of wood with curved longitudinal members, the strut ends being housed in steel sockets. The whole frame is braced with piano wire set diagonally between the struts. The car is floored from end to end, and the sides are enclosed with 8-ply wood covered with fabric.

Accommodation is provided for a wireless telegraphy operator, who is also a gunner, his compartment being situated forward, amidships is the pilot and abaft this seat is a compartment for the engineer.

The engine selected was the 75 horse-power water-cooled Rolls Royce, it being considered to be the most efficient for the purpose. The engine is mounted upon bearers above the level of the top of the car, and drives a four-bladed pusher propeller.

The car is suspended from an envelope of 70,000 cubic feet capacity, and the system of rigging is similar to that in use on all S.S. ships. The petrol is carried in aluminium tanks slung on the axis of the envelope, identically with the system in use on the S.S.P's. The usual elevator planes are adopted with a single long rudder plane.

The speed of the Zero is about 45 miles per hour and the ship has a theoretical endurance of seventeen hours; but this has been largely exceeded in practice.

The original ship proved an immediate success, and a large number was shortly afterwards ordered.

As time went on the stations expanded and sub-stations were added, while the Zero airship was turned out as fast as it could be built, until upwards of seventy had been commissioned. The work these ships were capable of exceeded the most sanguine expectations. Owing to their greater stability in flight and longer hours of endurance, they flew in weather never previously attempted by the earlier ships. With experience gained it was shown that a large fleet of airships of comparatively small capacity is of far more value for an anti-submarine campaign than a lesser fleet of ships of infinitely greater capacity. The average length of patrol was eight hours, but some wonderful duration flights were accomplished in the summer of 1918, as the following figures will show. The record is held by S.S.Z. 39, with 50 hours 55 minutes; another is 30 hours 20 minutes; while three more vary from 25 1/2 hours to 26 1/4. Although small, the Zero airship has been one of the successes of the war, and we can claim proudly that she is entirely a British product.

## S.S. TWIN

During the year 1917, designs were submitted for a twin-engined S.S. airship, the idea being to render the small type of airship less liable to loss from engine failure. The first design proved to be a failure, but the second was considered more promising, and several were built. Its capacity is 100,000 cubic feet, with a length of 164 feet 6 inches, and the greatest diameter 32 feet.

The car is built to carry five, with the engines disposed on gantries on the port and starboard side, driving pusher propellers. This type, although in the experimental stage, is being persevered with, and the intention is that it will gradually supplant the other S.S. classes. It is calculated that it will equal if not surpass the C Star ship in endurance, besides being easier to handle and certainly cheaper to build.

## "COASTAL" AND "C STAR" AIRSHIPS

The urgent need for a non-rigid airship to carry out anti-submarine patrol having been satisfied for the time with the production of the S.S. B.E. 2C type, the airship designers of the Royal Naval Air Service turned their attention to the production of an airship which would have greater lift and speed than the S.S. type, and, consequently, an augmented radius of action, together with a higher degree of reliability. As the name "Coastal" or "Coast Patrol" implies, this ship was intended to carry out extended sea patrols.

To obtain these main requirements the capacity of the envelope for this type was fixed at 170,000 cubic feet, as compared with the 60,000 cubic feet and, later, the 70,000 cubic feet envelopes adopted for the S.S. ships. Greater speed was aimed at by fitting two engines of 150 horse-power each, and it was hoped that the chances of loss owing to engine failure would be considerably minimized.

The Astra-Torres type of envelope, with its system of internal rigging, was selected for this class of airship; in the original ship the envelope used was that manufactured by the French Astra-Torres Company, and to which it had been intended to rig a small enclosed car. The ship in question was to be known as No. 10. This plan was, however, departed from, and the car was subsequently rigged to the envelope of the Eta, and a special car was designed and constructed for the original Coastal. Coastal airship No. 1 was commissioned towards the end of 1915 and was retained solely for experimental and training purposes. Approximately thirty of these airships were constructed during the year 1916, and were allocated to the various stations for patrol duties.

The work carried out by these ships during the two and a half years in which they were in commission, is worthy of the highest commendation. Before the advent of later and more reliable ships, the bulk of anti-submarine patrol on the east coast and south-west coast of England was maintained by the Coastal. On the east coast, with the prevailing westerly and south-westerly winds, these airships had many long and arduous voyages on their return from patrol, and in the bitterness of winter their difficulties were increased ten-fold. To the whole-hearted efforts of Coastal pilots and crews is due, to a great extent, the recognition which somewhat tardily was granted to the Airship Service.

The envelope of the Coastal airship has been shown to be of 170,000 cubic feet capacity. It is trilobe in section to employ the Astra-Torres system of internal and external rigging. The great feature of this principle is that it enables the car to be slung much closer to the envelope than would be possible with the tangential system on an envelope of this size. As a natural consequence there is far less head resistance, owing to the much shorter rigging, between the envelope and the car.

The shape of the envelope is not all that could have been desired, for it is by no means a true streamline, but has the same cross section for the greater part of its length, which tapers at either end to a point which is slightly more accentuated aft. Owing to the shape, these ships, in the early days until experience had been gained, were extremely difficult to handle, both on the landing ground and also in the air. They were extremely unstable both in a vertical and horizontal plane, and were slow in answering to their rudders and elevators.

The envelope is composed of rubber-proofed fabric doped to hold the gas and resist the effects of weather. Four ballonets are situated in the envelope, two in each of the lower lobes, air being conveyed to them by means of a fabric air duct, which is parallel to the longitudinal centre line of the envelope, with transverse ducts connecting each pair of ballonets. In earlier types of the Coastal, the air scoop supplying air to the air duct was fitted in the slip stream of the forward engine, but later this was fitted aft of the after engine.

Six valves in all are used, four air valves, one fitted to each ballonet, and two gas valves. These are situated well aft, one to each of the lower lobes, and are fitted on either side of the rudder plane. A top valve is dispensed with because in practice when an Astra-Torres envelope loses shape, the tendency is for the tail to be pulled upwards by the rigging, with the result that the two gas valves always remain operative.

Crabpots and non-return valves are employed in a similar manner to S.S. airships.

The Astra-Torres system of internal rigging must now be described in some detail. The envelope is made up of three longitudinal lobes, one above and two below, which when viewed end on gives it a trefoil appearance. The internal rigging is attached to the ridges formed on either side of the upper lobe, where it meets the two side lobes. From here it forms a V, when viewed cross sectionally, converging at he ridge formed by the two lobes on the underside of the envelope which is known as the lower ridge.

To the whole length of the top ridges are attached the internal rigging girdles and also the lacing girdles to which are secured the top and side curtains. These curtains are composed of ordinary unproofed fabric and their object is to make the envelope keep its trilobe shape. They do not, however, divide the ship into separate gas compartments. The rigging girdle consists of a number of fabric scallops through which run strands of Italian hemp. These strands, of which there are a large number, are led towards the bottom ridge, where they are drawn together and secured to a rigging sector. To these sectors the main external rigging cables are attached. The diagram shows better than any description this rigging system.

Ten main suspensions are incorporated in the Coastal envelope, of which three take the handling guys, the remaining seven support the weight of the car.

The horizontal fins with the elevator flaps, and the vertical fin with the rudder flap, are fixed to the ridges of the envelope.

The car was evolved in the first instance by cutting away the tail portion of two Avro seaplane fuselages and joining the forward portions end on, the resulting car, therefore, had engines at either end with seating accommodation for four. The landing chassis were altered, single skids being substituted for the wider landing chassis employed in the seaplane. The car consists of four longerons with struts vertical and cross, and stiffened with vertical and cross bracing wires. The sides are covered with fabric and the flooring and fairing on the top of the car are composed of three-ply wood. In the later cars five seats were provided to enable a second officer to be carried.

The engines are mounted on bearers at each end of the car, and the petrol and oil tanks were originally placed adjoining the engines in the car. At a later date various methods of carrying the petrol tanks were adopted, in some cases they were slung from the envelope and in others mounted on bearers above the engines.

Wireless telegraphy is fitted as is the case with all airships. In the Coastal a gun is mounted on the top of the envelope, which is reached by a climbing shaft passing through the envelope, another mounting being provided on the car itself.

Bombs are also carried on frames attached to the car. Sunbeam engines originally supplied the motive power, but at a later date a 220 horse-power Renault was fitted aft and a 100 horse-power, Berliet forward. With the greater engine power the ship's capabilities were considerably increased.

Exceedingly long flights were achieved by this type of ship, and those exceeding ten hours are far too numerous to mention. The most noteworthy of all gave a total of 24 1/4 hours, which, at the time, had only once been surpassed by any British airship.

Towards the end of 1917, these ships, having been in commission for over two years, were in many cases in need of a complete refit. Several were put in order, but it was decided that this policy should not be continued, and that as each ship was no longer fit for flying it should be replaced by the more modern Coastal known as the C Star.

The record of one of these ships so deleted is surely worthy of special mention. She was in commission for 2 years 75 days, and averaged for each day of this period 3 hours 6 minutes flying. During this time she covered upwards of 66,000 miles. From this it will be seen that she did not pass her life by any means in idleness.

## "C STAR" AIRSHIP

After considerable experience had been gained with the Coastal, it became obvious that a ship was required of greater capabilities to maintain the long hours of escort duty and also anti-submarine patrols. To meet these requirements it was felt that a ship could be constructed, not departing to any extent from the Coastal, with which many pilots were now quite familiar, but which would show appreciable improvement over its predecessor.

The design which was ultimately adopted was known as the C Star, and provided an envelope of 210,000 cubic feet, which secured an extra ton and a quarter in lifting capacity. This envelope, although of the Astra-Torres type, was of streamline form, and in that respect was a great advance on the early shape as used in the Coastal. It is to all intents and purposes the same envelope as is used on the North Sea ships, but on a smaller scale. An entirely new type of fabric was employed for this purpose. The same model of car was employed, but was made more comfortable, the canvas covering for the sides being replaced by three-ply wood. In all other details the car remained entirely the same. The standard power units were a 100 horse-power Berliet forward and a Fiat of 260 horse-power aft. The petrol tanks in this design were carried inside the envelope, which was quite a new departure.

These airships may be considered to have been successful, though not perhaps to the extent which was expected by their most ardent admirers. With the advent of the S.S. Twin it was resolved not to embark on a large constructional programme, and when the numbers reached double figures they were no longer proceeded with. Notwithstanding this the ships which were commissioned carried out most valuable work, and, like their prototypes, many fine flights were recorded to their credit. Thirty-four and a half hours was the record flight for this type of ship, and another but little inferior was thirty hours ten minutes. These flights speak well for the endurance of the crews, as it must be borne in mind that no sleeping accommodation is possible in so small a car.

The Coastal airship played no small part in the defeat of the submarine, but its task was onerous and the enemy and the elements unfortunately exacted a heavy toll. A German wireless message received in this country testified to the valiant manner in which one of these ships met with destruction.

## THE "NORTH SEA" AIRSHIP

The North Sea or N.S. airship was originally designed to act as a substitute for the Rigid, which, in 1916, was still a long way from being available for work of practical utility. From experience gained at this time with airships of the Coastal type it was thought possible to construct a large Non-Rigid capable of carrying out flights of twenty-four hours' duration, with a speed of 55 to 60 knots, with sufficient accommodation for a double crew.

The main requirements fall under four headings:

1. Capability to carry out flights of considerable duration.

2. Great reliability.

3. The necessary lift to carry an ample supply of fuel.

4. Adequate arrangements to accommodate the crew in comfort.


If these could be fulfilled the authorities were satisfied that ships possessing these qualifications would be of value to the Fleet and would prove efficient substitutes until rigid airships were available. The North Sea, as may be gathered from its name, was intended to operate on the east coasts of these islands.

The first ship, when completed and put through her trials, was voted a success, and the others building were rapidly pushed on with. When several were finished and experience had been gained, after long flights had been carried out, the North Sea airship suffered a partial eclipse and people were inclined to reconsider their favourable opinion. Thus it was that for many months the North Sea airship was decidedly unpopular, and it was quite a common matter to hear her described as a complete failure. The main cause of the prejudice was the unsatisfactory design of the propelling machinery, which it will be seen later was modified altogether, and coupled with other improvements turned a ship of doubtful value into one that can only be commended.

The envelope is of 360,000 cubic feet capacity, and is designed on the Astra-Torres principle for the same reasons as held good in the cases of the Coastal and C Star. All the improvements which had been suggested by the ships of that class were incorporated in the new design, which was of streamline shape throughout, and looked at in elevation resembled in shape that of the S.S. airship. Six ballonets are fitted, of which the total capacity is 128,000 cubic feet, equivalent to 35.5 per cent of the total volume. They are fitted with crabpots and non-return valves in the usual manner.

The rigging is of the Astra-Torres system, and in no way differs from that explained in the previous chapter. Nine fans of the internal rigging support the main suspensions of the car, while similar fans both fore and aft provide attachment for the handling guys. Auxiliary fans on the same principle support the petrol tanks and ballast bag.

Four gas and six air valves in all are fitted, all of which are automatic.

Two ripping panels are embodied in the top lobe of the envelope.

The N.S. ship carries four fins, to three of which are attached the elevator and rudder flaps. The fourth, the top fin, is merely for stabilizing purposes, the other three being identical in design, and are fitted with the ordinary system of wiring and kingposts to prevent warping.

The petrol was originally carried in aluminium tanks disposed above the top ridges of the envelope, but this system was abandoned owing to the aluminium supply pipes becoming fractured as the envelope changed shape at different pressures. They were then placed inside the envelope, and this rearrangement has given every satisfaction.

To the envelope of the N.S. is rigged a long covered-in car. The framework of this is built up of light steel tubes, the rectangular transverse frames of which are connected by longitudinal tubes, the whole structure being braced by diagonal wires. The car, which tapers towards the stern, has a length of 85 feet, with a height of 6 feet. The forward portion is covered with duralumin sheeting, and the remainder with fabric laced to the framework. Windows and portholes afford the crew both light and space to see all that is required. In the forward portion of the car are disposed all the controls and navigating instruments, together with engine-telegraphs and voice pipes. Aft is the wireless telegraphy cabin and sleeping accommodation for the crew.

A complete electrical installation is carried of two dynamos and batteries for lights, signalling lamps, telephones, etc. The engines are mounted in a power unit structure separate from the car and reached by a wooden gangway supported by wire cables. This structure consists of two V-shaped frameworks connected by a central frame and by an under-structure to which floats are attached. The mechanics' compartment is built upon the central frame, and the engine controls are operated from this cabin.

In the original power units two 250 horse-power Rolls Royce engines were fitted, driving propellers on independent shafts through an elaborate system of transmission. This proved to be a great source of weakness, as continual trouble was experienced with this method, and a fracture sooner or later occurred at the universal joint nearest to the propeller. When the modified form of ship was built the whole system of transmission was changed, and the propellers were fitted directly on to the engine crankshafts.

At a later date 240 horse-power Fiat engines were installed, and the engineers' cabin was modified and an auxiliary blower was fitted to supply air to the ballonets for use if the engines are not running.

In the N.S. ship as modified the car has been raised to the same level as the engineers' cabin, and all excrescences on the envelope were placed inside. This, added to the improvement effected by the abolition of the transmission shafts, increased the reliability and speed of the ship, and also caused a reduction in weight.

The leading dimensions of the ship are as follows: length, 262 feet; width, 56 feet 9 inches; height, 69 feet 3 inches. The gross lift is 24,300 lb.; the disposable lift, without crew, petrol, oil, and ballast, 8,500 lb. The normal crew carried when on patrol is ten, which includes officers.

As in the case of the Coastal, a gun is mounted on the top of the envelope, which is approached by a similar climbing shaft, and guns and bombs are carried on the car.

These ships have become notorious for breaking all flying records for non-rigid airships. Even the first ship of the class, despite the unsatisfactory power units, so long ago as in the summer of 1917 completed a flight of 49 hours 22 minutes, which at the time was the record flight of any British airship. Since that date numerous flights of quite unprecedented duration have been achieved, one of 61 1/2 hours being particularly noteworthy, and those of upwards of 30 hours have become quite commonplace.

Since the Armistice one of these ships completed the unparalleled total of 101 hours, which at that date was the world's record flight, and afforded considerable evidence as to the utility of the non-rigid type for overseas patrol, and even opens up the possibility of employing ships of similar or slightly greater dimensions for commercial purposes.

N.S. 6 appeared several times over London in the summer months of 1918, and one could not help being struck by the ease with which she was steered and her power to remain almost stationary over such a small area as Trafalgar Square for a quite considerable period.

The flights referred to above were not in any way stunt performances to pile up a handsome aggregate of hours, but were the ordinary flying routine of the station to which the ships were attached, and most of the hours were spent in escorting convoys and hunting for submarines. In addition to these duties, manoeuvres were carried out on occasions with the Fleet or units thereof.

From the foregoing observations it must be manifest that this type of ship, in its present modified state, is a signal success, and is probably the best large non-rigid airship that has been produced in any country.

For the purposes of comparison it will be interesting to tabulate the performances of the standard types of non-rigid airships. The leading dimensions are also included in this summary:

| Type | S.S. Zero | S.S. Twin | Coastal | North Star Sea |
|---|---|---|---|---|
| Length | 143' 0" | 165' 0" | 218' 0" | 262' 0" |
| Overall width | 32' 0" | 35' 6" | 49' 3" | 56' 9" |
| Overall height | 46' 0" | 49' 0" | 57' 6" | 69' 3" |
| Hydrogen capacity (cubic feet) | 70,000 | 100,000 | 210,000 | 360,000 |
| Gross lift (lb.) | 4,900 | 7,000 | 14,500 | 24,300 |
| Disposable lift (lb.) | 1,850 | 2,200 | 4,850 | 8,500 |
| Crew | 3 | 4 | 5 | 10 |
| Lift available for fuel and freight (lb.) | 1,370 | 1,540 | 4,050 | 6,900 |
| Petrol consumption at full speed (lb. per hour) | 3.6 | 7.2 | 18.4 | 29.8 |
| Gals. per hour | 0.36 | 0.72 | 2.05 | 3 |

# CHAPTER VII

## NAVAL AIRSHIPS.--THE RIGIDS--RIGID AIRSHIP No. 1

The responsibility for the development the Rigid airship having been allotted to the Navy, with this object in view, in the years 1908 and 1909 a design was prepared by Messrs. Vickers Ltd., in conjunction with certain naval officers, for a purely experimental airship which should be as cheap as possible. The ship was to be known as Naval Airship No. 1, and though popularly called the Mayfly, this title was in no way official. In design the following main objects were aimed at:

1. The airship was to be capable of carrying out the duties of an aerial scout.
2. She was to be able to maintain a speed of 40 knots for twenty-four hours, if possible.
3. She was to be so designed that mooring to a mast on the water was to be feasible, to enable her to be independent of her shed except for docking purposes, as in the case with surface vessels.

4. She was to be fitted with wireless telegraphy.

5. Arrangements were to be made for the accommodation of the crew in reasonable comfort.
6. She was to be capable of ascending to a height of not less than 1,500 feet.

These conditions rendered it necessary that the airship should be of greater dimensions than any built at the time, together with larger horse-power, etc.

These stipulations having been settled by the Admiralty, the Admiralty officials, in conjunction with Messrs. Vickers Ltd., determined the size, shape, and materials for the airship required. The length of the ship was fixed at approximately 500 feet, with a diameter of 48 feet. Various shapes were considered, and the one adopted was that recommended by an American professor named Zahm. In this shape, a great proportion of the longitudinal huff framework is parallel sided with curved bow and stern portions, the radius of these curved portions being, in the case of the bow, twice the diameter of the hull, and in the case of the stern nine times the same diameter. Experiments proved that the resistance of a ship of this shape was only two-fifths of the resistance of a ship of the same dimensions, having the 1 1/2 calibre bow and stern of the Zeppelin airships at that time constructed.

A considerable difference of opinion existed as to the material to be chosen for the construction of the hull. Bamboo, wood, aluminium, or one of its alloys, were all considered. The first was rejected as unreliable. The second would have been much stronger than aluminium, and was urged by Messrs. Vickers. The Admiralty, however, considered that there was a certainty of better alloys being produced, and as the ship was regarded as an experiment and its value would be largely negatived if later ships were constructed of a totally different material, aluminium or an alloy was selected. The various alloys then in existence showed little advantage over the pure metal, so pure aluminium was specified and ordered. This metal was expected to have a strength of ten tons per square inch, but that which arrived was found to be very unreliable, and many sections had, on test, only half the strength required. The aluminium wire intended for the mesh wiring of the framework was also found to be extremely brittle. A section of the framework was, however, erected, and also one of wood, as a test for providing comparisons. In the tests, the wooden sections proved, beyond all comparison, the better, but the Admiralty persisted in their decision to adopt the metal.

Towards the end of 1909 a new aluminium alloy was discovered, known as duralumin. Tests were made which proved that this new metal possessed a strength of twenty-five tons per square inch, which was over twice as strong as the nominal strength of aluminium, and in practice was really five times stronger. The specific gravity of the new metal varied from 2.75 to 2.86, as opposed to the 2.56 of aluminium. As the weights were not much different it was possible to double the strength of the ship and save one ton in weight. Duralumin was therefore at once adopted.

The hull structure was composed of twelve longitudinal duralumin girders which ran fore and aft the length of the ship and followed the external shape. The girders were secured to a steel nose-piece at the bow and a pointed stern-piece aft. These girders, built of duralumin sections, were additionally braced wherever the greatest weights occurred. To support these girders in a thwartship direction a series of transverse frames were placed at 12 feet 6 inches centres throughout the length of the ship, and formed, when viewed cross-sectionally, a universal polygon of twelve sides. For bracing purposes mesh wiring stiffened each bay longitudinally, so formed by the junction of the running girder and the transverse frames, while the transverse frames between the gasbags were stiffened with radial wiring which formed structure similar to a wheel with its spokes. The frames where the gondolas occurred were strengthened to take the addition weight, while the longitudinals were also stiffened at the bow and stern.

Communication was provided between the gondolas by means of an external keel which was suspended from extra keel longitudinals. In this design the keel was provided for accommodation purposes only, and in no way increased the structural stability of the ship as in No. 9 and later ships. This keel, triangular in section, widened out amidships to form a space for a cabin and the wireless compartment. The fins and rudders, which were adopted, were based entirely on submarine experience, and the Zeppelin method was ignored. The fins were fitted at the stern of the ship only, and comprised port and starboard horizontal fins, which followed approximately the shape of the hull, and an upper and lower vertical fin. Attached to these fins were box rudders and elevators, instead of the balanced rudders first proposed. Auxiliary rudders were also fitted in case of a breakdown of the main steering gear abaft the after gondola. Elevators and rudders were controlled from the forward gondola and the auxiliary rudders from the after gondola.

The gasbags were seventeen in number and were twelve-sided in section, giving approximately a volume of 663,000 cubic feet when completely full. Continental fabric, as in use on the Zeppelin airships, was adopted, although the original intention was to use goldbeater's skin, but this was abandoned owing to shortage of material. These bags were fitted with the Parseval type of valve, which is situated at the top, contrary to the current Zeppelin practice, which had automatic valves at the bottom of the bags, and hand-operated valves on the top of a few bags for control purposes. Nets were laced to the framework to prevent the bags bulging through the girders.

The whole exterior of the hull was fitted with an outer cover; Zeppelin at this time used a plain light rubber-proofed fabric, but this was not considered suitable for a ship which was required to be moored in the open, as in wet weather the material would get saturated and water-logged. Various experiments were carried out with cotton, silk and ramie, and, as a result, silk treated with Ioco was finally selected. This cover was laced with cords to the girder work, and cover-strips rendered the whole impervious to wet. Fireproofed fabric was fitted in wake of the gondolas for safety from the heat of the engines.

Two gondolas, each comprising a control compartment and engine-room, were suspended from the main framework of the hull. They were shaped to afford the least resistance possible to the air, and were made of Honduras mahogany, three-ply where the ballast tanks occurred, and two-ply elsewhere. The plies were sewn together with copper wire. The gondolas were designed to have sufficient strength to withstand the strain of alighting on the water. They were suspended from the hull by wooden struts streamline in shape, and fitted with internal steel-wire ropes; additional wire suspensions were also fitted to distribute the load over a greater length of the ship. The engines were carried in the gondolas on four hollow wooden struts, also fitted internally with wire. The wires were intended to support the gondolas in the event of the struts being broken in making a heavy landing.

Two engines were mounted, one in each gondola, the type used being the 8-cylinder vertical water-cooled Wolseley developing a horse-power of 160. The forward engine drove two wing propellers through the medium of bevel gearing, while the after engine drove a single large propeller aft through 4 gear box to reduce the propeller revolutions to half that of the engine. The estimated speed of the ship was calculated to be 42 miles per hour, petrol was carried in tanks, fitted in the keel, and the water ballast tanks were placed close to the keel and connected together by means of a pipe.

No. 1 was completed in May, 1911. She had been built at Barrow in a shed erected on the edge of Cavendish Dock. Arrangements were made that she should be towed out of the shed to test her efficiency at a mooring post which had been prepared in the middle of the dock. She was launched on May 22nd in a flat calm and was warped out of the shed and hauled to the post where she was secured without incident. The ship rode at the mooring post in a steady wind, which at one time increased to 36 miles per hour, until the afternoon of May 25th, and sustained no damage whatever. Various engine trials were carried out, but no attempt was made to fly, as owing to various reasons the ship was short of lift. Valuable information was, however, gained in handling the ship, and much was learnt of her behaviour at the mast. More trouble was experienced in getting her back into the shed, but she was eventually housed without sustaining any damage of importance.

Owing to the lack of disposable lift, the bags were deflated and various modifications were carried out to lighten the ship, of which the principal were the removal of the keel and cabin entirely, and the removal of the water-trimming services. Other minor alterations were made which gave the ship, on completion, a disposable lift of 3.21 tons. The transverse frames between the gasbags were strengthened, and a number of broken wires were replaced.

On September 22nd the ship was again completed, and on the 24th she was again to be taken out and tested at the mooring post. Unfortunately, while being hauled across the dock, the framework of the ship collapsed, and she was got back into the shed the same day.

Examination showed that it was hopeless to attempt to reconstruct her, and she was broken up at a later date. The failure of this ship was a most regrettable incident, and increased the prejudice against the rigid airship to such an extent that for some time the Navy refused to entertain any idea of attempting a second experiment.

## RIGID AIRSHIP No. 9

Rigid Airship No. 1 having met with such a calamitous end, the authorities became rather dubious as to the wisdom of continuing such costly experiments. Most unfortunately, as the future showed and as was the opinion of many at the time, rigid construction in the following year 1912 was ordered to be discontinued. This decision coincided with the disbanding of the Naval Air Service, and for a time rigid airships in this country were consigned to the limbo of forgetfulness. After the Naval Air Service had been reconstituted, the success which attended the Zeppelin airships in Germany could no longer be overlooked, and it was decided to make another attempt to build a rigid airship in conformity with existing Zeppelin construction. The first proposals were put forward in 1913, and, finally, after eleven months delay, the contract was signed. This airship, it has been seen, was designated No. 9.

No. 9 experienced numerous vicissitudes, during the process of design and later when construction was in progress. The contract having been signed in March, 1914, work on the ship was suspended in the following February, and was not recommenced until July of the same year. From that date onwards construction was carried forward; but so many alterations were made that it was fully eighteen months before the ship was completed and finally accepted by the Admiralty.

The ship as designed was intended "to be generally in conformity with existing Zeppelin construction," with the following main requirements stipulated for in the specification:

1. She was to attain a speed of at least 45 miles per hour at the full power of the engines.
2. A minimum disposable lift of five tons was to be available for movable weights.
3. She was to be capable of rising to a height of 2,000 feet during flight.

The design of this ship was prepared by Messrs. Vickers, Ltd., and as it was considered likely that owing to inexperience the ship would probably be roughly handled and that heavy landings might be made, it was considered that the keel structure and also the cars should be made very strong in case of accidents occurring. This, while materially increasing the strength of the ship, added to its weight, and coupled with the fact that modifications were made in the design, rendered the lift somewhat disappointing. The hull structure was of the "Zahm" shape as in No. 1, a considerable portion being parallel sided, while in transverse section it formed a 17-sided polygon. In length it was 526 feet with a maximum diameter of 53 feet. The hull framework was composed of triangular duralumin girders, both in the longitudinal and transverse frames, while the bracing was carried out by means of high tensile steel wires and duralumin tubes. Attached to the hull was a V-shaped keel composed of tubes with suitable wire bracings, and in it a greater part of the strength of the structure lay. It was designed to withstand the vertical forces and bending moments which resulted from the lift given by the gasbags and the weights of the car and the cabin. The keel also provided the walking way from end to end of the ship, and amidships was widened out to form a cabin and wireless compartment.

The wiring of the transverse frames was radial and performed similar functions to the spokes of a bicycle wheel. These wires could be tightened up at the centre at a steel ring through which they were threaded and secured by nuts.

In addition to the radial wires were the lift wires, which were led to the two points on the transverse frames which were attached to the keel; on the inflation of the gasbags, the bags themselves pressed upon the longitudinal girders on the top of the ship, which pressure was transferred to the transverse frames and thence by means of the several lift wires to the keel. In this way all the stresses set up by the gas were brought finally to the keel in which we have already said lay the main strength of the ship.

The hull was divided by the transverse frames into seventeen compartments each containing a single gasbag. The bags were composed of rubber-proofed fabric lined with gold-beater's skin to reduce permeability, and when completely full gave a total volume of 890,000 cubic feet. Two types of valve were fitted to each bag, one the Parseval type of valve with the pressure cone as fitted in No. 1, the other automatic but also controlled by hand.

To distribute the pressure evenly throughout the upper longitudinal frames, and also to prevent the gasbags bulging between the girders, nets were fitted throughout the whole structure of the hull.

The whole exterior of the ship was fitted with an outer cover, to protect the gasbags and hull framework from weather and to render the outer surface of the ship symmetrical and reduce "skin friction" and resistance to the air to a minimum. To enable this cover to be easily removed it was made in two sections, a port and starboard side for each gasbag. The covers were laced to the hull framework and the connections were covered over with sealing strips to render the whole weathertight.

The system of fins for stabilizing purposes on No. 9 were two--vertical and horizontal. The vertical fin was composed of two parts, one above and the other below the centre line of the ship.

They were constructed of a framework of duralumin girders, covered over with fabric. The fins were attached on one edge to the hull structure and wire braced from the other edge to various positions on the hull. The horizontal fins were of similar design and attached in a like manner to the hull. Triplane rudders and biplane elevators of the box type were fitted in accordance with the German practice of the time. Auxiliary biplane rudders were fitted originally abaft the after car, but during the first two trial flights they proved so very unsatisfactory that it was decided to remove them.

Two cars or gondolas were provided to act as navigating compartments and a housing for the engines, and in design were calculated to offer the least amount of head resistance to the wind. The cars were composed of duralumin girders, which formed a flooring, a main girder running the full length of the car with a series of transverse girders spaced in accordance with the main loads. From each of these transverse girders vertical standards with a connecting piece on top were taken and the whole exterior was covered with duralumin plating. The cars were suspended in the following manner. Two steel tubes fitting into a junction piece at each end were bolted to brackets at the floor level at each end of the transverse girders. They met at an apex above the roof level and were connected to the tubing of the keel. In addition, to distribute the weight and prevent the cars from rocking, steel wire suspensions were led to certain fixed points in the hull.

Each car was divided into two parts by a bulkhead, the forward portion being the control compartment in which were disposed all instruments, valve and ballast controls, and all the steering and elevating arrangements. Engine-room telegraphs, voice pipes and telephones were fitted up for communication from one part of the ship to the other. The keel could be reached by a ladder from each car, thus providing with the climbing shaft through the hull access to all parts of the ship.

The original engine equipment of No. 9 was composed of four Wolseley-Maybach engines of 180 horse-power each, two being installed in the forward car and two in the after car. As the ship was deficient in lift after the initial flight trials had been carried out, it was decided to remove the two engines from the after car and replace them with a single engine of 250 horse-power; secondly, to remove the swivelling propeller gear from the after car and substitute one directly-driven propeller astern of the car. This as anticipated reduced the weight very considerably and in no way lessened the speed of the ship.

The forward engines drove two four-bladed swivelling propellers through gear boxes and transmission shafts, the whole system being somewhat complicated, and was opposed to the Zeppelin practice at the time which employed fixed propellers.

The after engine drove a large two-bladed propeller direct off the main shaft.

The petrol and water ballast were carried in tanks situated in the keel and the oil was carried in tanks beneath the floors of the cars.

The wireless cabin was situated as before mentioned in a cabin in the keel of the ship, and the plant comprised a main transmitter, an auxiliary transmitter and receiver and the necessary aerial for radiating and receiving.

No. 9 was inflated in the closing days of 1916, and the disposal lift was found to be 2.1 tons under the specification conditions, namely, barometer 29.5 inches and temperature 55 degrees Fahrenheit. The contract requirements had been dropped to 3.1 tons, which showed that the ship was short by one ton of the lift demanded. The flight trials were, however, carried out, which showed that the ship had a speed of about 42 1/2 miles per hour.

The alterations previously mentioned were afterwards made, the bags of the ship were changed and another lift and trim trial was held in March, 1917, when it was found that these had had the satisfactory result of increasing the disposable lift to 3.8 tons or .7 ton above the contract requirements, and with the bags 100 per cent full gave a total disposable lift of 5.1 tons.

Additional trials were then carried out, which showed that the speed of the ship had not been impaired.

For reference purposes the performances of the ship are tabulated below.

>Speed: Full 45 miles per hour Normal = 2/3 38 " " " Cruising = 1/3 32 " " " Endurance: Full 18 hours = 800 miles Normal 26 " = 1,000 " Cruising 50 " = 1,600 "

No. 9 having finished her trials was accepted by the Admiralty in Mar. 1917, and left Barrow, where she had been built, for a patrol station.

In many ways she was an excellent ship, for it must be remembered that when completed she was some years out-of-date judged by Zeppelin standards. Apart from the patrol and convoy work which she accomplished, she proved simply invaluable for the training of officers and men selected to be the crews of future rigid airships. Many of these received their initial training in her, and there were few officers or men in the airship service who were not filled with regret when orders were issued that she was to be broken up. The general feeling was that she should have been preserved as a lasting exhibition of the infancy of the airship service, but unfortunately rigid airships occupy so much space that there is no museum in the country which could have accommodated her. So she passed, and, except for minor trophies, remains merely a recollection.

## RIGID AIRSHIP No. 23 CLASS

After the decision had been made in 1915 that work on No. 9 should be restarted, the Admiralty determined that a programme of rigid airships should be embarked upon, and design was commenced.

Several ships of the same class were, ordered, and the type was to be known as the 23 class. Progress on these ships, although slow, was more rapid than had been the case with No. 9, and by the end of 1917 three were completed and a fourth was rapidly approaching that state.

The specification, always ambitious, laid down the following main stipulations.

(1) The ship is to attain a speed of at least 55 miles per hour for the main power of the engines.
(2) A minimum of 8 tons is to be available for disposable weights when full.
(3) The ship must be capable of rising at an average rate of not less than 1,000 feet per minute, through a height of 3,000
feet starting from nearly sea level.

As will be seen later this class of ship, although marking a certain advance on No. 9 both as regards workmanship and design, proved on the whole somewhat disappointing, and it became more evident every day that we had allowed the Germans to obtain such a start in the race of airship construction as we could ill afford to concede.

We may here state that all of the ships of this class which had been ordered were not completed, the later numbers being modified into what was known as the 23 X class; four in all of the 23 class were built, of which two--Nos. 23 and 26--were built by Messrs. Vickers, Ltd., at Barrow, No. 24 by Messrs. Wm. Beardmore and Co., at Glasgow, and No. 25 by Messrs. Armstrong, Whitworth and Co., at Selby, Yorkshire.

In many respects the closest similarity of design exists between No. 9 and No. 23, especially in the hull, but it will be of interest to mention the salient differences between the two ships.

The length of the hull, which in No. 9 was 520 feet, was increased in No. 23 to 535 feet, and the number of gasbags from seventeen to eighteen. This gave a total volume of 997,500 cubic feet compared with 890,000 cubic feet in No. 9, with a disposable lift under specification conditions of 5.7 tons as opposed to 3.8 tons.

The longitudinal shape of No. 23 is a modified form of "Zahm" shape, the radius of the bow portion being twice the diameter of the parallel portion, while the stern radius is three times the same diameter.

In design the hull framework is almost a repetition of No. 9, particularly in the parallel portion, the same longitudinal and transverse frames dividing the hull into compartments, with tubes completely encircling the section between each main transverse frame. The system of wiring the hull is precisely the same in both the ships, and nets are employed in the same way.

The triangular section of keel is adhered to, but its functions in No. 23 are somewhat different. In No. 9 it was intended to be sufficiently strong to support all the main vertical bending moments and shearing forces, but in No. 23 it was primarily intended to support the distributed weights of water ballast, petrol tanks, etc., between the main transverse frames. Unlike No. 9, the keel is attached to the main transverse frames only. The cabin and wireless cabin are disposed in the keel in the same manner, and it also furnishes a walking way for the total length of the ship.

The stabilizing fins, both vertical and horizontal, are similar to those attached to No. 9, but the system of rudders and elevators is totally different. In place of the box rudders and elevators in No. 9, single balanced rudders and elevators are attached to the fins; they have their bearing on the outboard side on the external girders of the fins, which are extended for the purpose. The elevators and rudders are composed of a duralumin framework, stiffened by a kingpost on either side with bracing wires.

The bags, eighteen in number, are made of rubber-proofed fabric lined with gold-beater's skin. It is interesting to note that the number of skins used for the bags of a ship of this class is approximately 350,000. The system of valves is entirely different from that in No. 9. The Parseval type of valve with the pressure cone at the bottom of the bag is omitted, and in the place of the two top valves in the former ship are a side valve of the Zeppelin type entirely automatic and a top valve entirely hand controlled. The side valve is set to blow off at a pressure of from 3 to 5 millimetres. The outer cover was fitted in the same manner as in No. 9. Two cars or gondolas, one forward, the other aft, each carry one engine provided with swivelling propellers and gears. They are enclosed with sides and a fireproof roof, and are divided into two compartments, one the navigating compartment, the other the engine room. The cars are in all respects very similar to those of No. 9, and are suspended from the hull in a similar manner. The remaining two engines are carried in a small streamline car situated amidships, which has just sufficient room in it for the mechanics to attend to them. Originally this car was open at the top, but it was found that the engineers suffered from exposure, and it was afterwards roofed in.

The engine arrangements in this ship were totally different to those of No. 9, four 250 horse-power Rolls Royce engines being installed in the following order. Single engines are fitted in both the forward and after cars, each driving two swivelling four-bladed propellers. In the centre car two similar engines are placed transversely, which drive single fixed propellers mounted on steel tube outriggers through suitable gearing.

The engines are the standard 12 cylinder V-type Rolls Royce which will develop over 300 brake horse-power at full throttle opening.

The engine is water cooled, and in the case of those in the forward and after cars the original system consisted of an internal radiator supplied by an auxiliary water tank carried in the keel. It was found on the flight trials that the cooling was insufficient, and external radiators were fitted, the internal radiator and fan being removed. In the case of the centre car no alteration was necessary, as external radiators were fitted in the first instance.

The engines are supported by two steel tubes held by four brackets bolted to the crank case, these being carried by twelve duralumin tubes bolted to the bearers and transverse frames of the car respectively. The drive from the engine is transmitted through a universal joint to a short longitudinal shaft, running on ball bearings. This shaft gears into two transverse shafts, which drive the propellers through the medium of a gear box to the propeller shafts, making five shafts in all.

The engines in the centre car being placed transversely the transmission is more direct, the engines driving the propellers through two gear wheels only. The propeller gear box is supported by steel tube outriggers attached by brackets to the framework of the car. The petrol is carried in a series of tanks situated beneath the keel walking way, and are interconnected so that any tank either forward or aft can supply any engine, by this means affording assistance for the trimming of the ship.

Four-bladed propellers are used throughout the ship.

Water ballast is carried in fabric bags also situated beneath the keel walking way, and a certain amount is also carried beneath the floor of the car.

Engine-room telegraphs, swivelling propeller telegraphs, speaking tubes and telephones, with a lighting set for the illumination of the cars and keel, were all fitted in accordance with the practice standard in all rigid airships.

The lift and trim trials taken before the initial flight trials showed that the ship possessed a disposable lift under standard conditions of 5.7 tons. The original disposable lift demanded by the specification was 3 tons but this was reduced by 2 tons owing to the machinery weights being 2 tons in excess of the estimate. Since then these weights had been increased by another half-ton, making a total of 2 1/2 tons over the original estimate. It was evident that with so small a margin of lift these ships would never be of real use, and it was decided to remove various weights to increase the lift and to substitute a wing car of a similar type to those manufactured for the R 33 class for the heavy after car at present in use.

R 23 carried out her trials without the alteration to the car, which was effected at a later date, and the same procedure was adopted with R 24 and R 25. In the case of R 26, however, she had not reached the same stage of completion as the other two ships, and the alterations proposed for them were embodied in her during construction. The gasbags were of lighter composition, all cabin furniture was omitted and the wing car was fitted in place of the original after car. This wing car is of streamline shape with a rounded bow and tapered stern. The lower portion is plated with duralumin sheets and the upper part is covered with canvas attached to light wooden battens to give the necessary shape. This effected a very considerable reduction in weight. The original 250 horsepower Rolls Royce engine was installed, now driving a single large two-bladed propeller astern. A test having been taken, it was found that the disposable lift under standard conditions was 6.28 tons. It was therefore decided that all the ships of the class should be modified to this design when circumstances permitted. Speed trials were carried out under various conditions of running, when it was found that the ship possessed a speed of 54 1/4 miles per hour with the engines running full out.

To summarize the performances of these ships as we did in the case of No. 9, we find:

> Speed: Full 54 miles per hour Normal =2/3 48 " " " Cruising =1/3 33 " " " Endurance: Full 18 hours = 1,000 miles Normal 26 " = 1,250 " Cruising 50 " = 1,900 "

The production of the rigid airship during the war was always surrounded with a cloak of impenetrable mystery. Few people, except those employed on their construction or who happened to live in the immediate vicinity of where they were built, even knew of their existence, and such ignorance prevailed concerning airships of every description that the man in the street hailed a small non-rigid as "the British Zeppelin" or admired the appearance of R 23 as "the Silver Queen." The authorities no doubt knew their own business in fostering this ignorance, although for many reasons it was unfortunate that public interest was not stimulated to a greater degree. In the summer months of 1918, however, they relented to a certain extent, and R 23 and one of her sister ships were permitted to make several flights over London to the intense delight of thousands of its inhabitants, and a certain amount of descriptive matter appeared in the Press.

From that time onwards these large airships have completely captured the popular imagination, and many absurd rumours and exaggerations have been circulated regarding their capabilities. It has been gravely stated that these airships could accomplish the circuit of the globe and perform other feats of the imagination. It must be confessed that their merits do not warrant these extravagant assertions. The fact remains, however, that R 23 and her sister ship R 26 have each carried out patrols of upwards of 40 hours duration and that, similarly to No. 9, they have proved of the greatest value for training airship crews and providing experience and data for the building programme of the future. At the present time highly interesting experiments are being carried out with them to determine the most efficient system of mooring in the open, which will be discussed at some length in the chapter dealing with the airship of the future.

## RIGID AIRSHIP 23 X CLASS

During the early days of building the airships of the 23 class, further information was obtained relating to rigid airship construction in Germany, which caused our designers to modify their views. It was considered a wrong policy to continue the production of a fleet of ships the design of which was becoming obsolete, and accordingly within ten months of placing the order for this class a decision was reached that the last four ships were to be altered to a modified design known as the 23 X class. As was the case with the ships of the preceding class when nearing completion, they were realized to be out of date, and special efforts being required to complete the ships of the 33 class and to release building space for additional larger ships, the construction of the second pair was abandoned.

The main modification in design was the abolition of the external keel, and in this the later Zeppelin principles were adopted. This secured a very considerable reduction in structural weight with a corresponding large expansion of the effective capabilities of the ship.

It has been seen that the purpose of the keel in No. 9 was to provide a structure sufficiently strong to support all the main vertical bending moments and shearing forces, and that in No. 23 this principle was somewhat different, in that the keel in this ship was primarily intended to support the distributed weights of petrol, water, ballast, etc., between the transverse frames.

In this later design, namely, the 23 X class, it was considered that the weights could be concentrated and suspended from the radial wiring of the transverse frames and that the keel, incorporated in the design of the former ships, could be dispensed with.

For all practical purposes, apart from the absence of the keel, the 23 X class of airship may be regarded as a slightly varied model of the 23 class. The main dimensions are nearly the same, and the general arrangement of the ship is but little changed. The loss of space owing to the introduction of the internal corridor is compensated by a modification of the shape of the bow, which was redesigned with a deeper curve. The hull structure was also strengthened by utilizing a stronger type of girder wherever the greatest weights occur. In these strengthened transverse frames the girders, while still remaining of the triangular section, familiar in the other ships, are placed the opposite way round, that is, with the apex pointing outwards.

The walking way is situated at the base of the hull passing through the gas chambers, which are specially shaped for the purpose. The corridor is formed of a light construction of hollow wooden struts and duralumin arches covered with netting.

In all other leading features the design of the 23 class is adhered to; the gasbags are the same, except for the alteration due to the internal corridor, and the system of valves and the various controls are all highly similar.

The arrangement of gondolas and the fitting of engines in all ways corresponds to the original arrangement of R 23, with the exception that they were suspended closer to the hull owing to the absence of the external keel. The substitution of the wing car of the 33 class for the original after gondola, carried out in the modifications undergone by the ships of the 23 class, was not adopted in these ships, as the wireless compartment installed in the keel in the former was fitted in the after gondola in the latter.

The disposable lift of these ships under standard conditions is 7 1/2 tons, which shows considerable improvement on the ships of the former classes.

Summarizing as before, the performances appear as under--

Speed: Full 56 1/2 miles per hour Normal 53 " " " Cruising 45 " " " Endurance: Normal 19 hours = 1,015 miles Cruising 23 1/2 " = 1,050 "

The two ships of this class, which were commissioned, must be regarded within certain limits as most satisfactory, and are the most successful of those that appeared and were employed during the war. Escort of convoys and extended anti-submarine patrols were carried out, and certain valuable experiments will be attempted now that peace has arrived.

In spite of the grave misgivings of many critics, the structure without the keel has proved amply strong, and no mishap attended this radical departure on the part of the designers.

## RIGID AIRSHIP No. 31 CLASS

The airship known as R 81 was a complete deviation from any rigid airship previously built in this country. In this case the experiment was tried of constructing it in wood in accordance with the practice adopted by the Schutte-Lanz Company in Germany.

It must be frankly acknowledged that this experiment resulted in failure. The ship when completed showed great improvement both in shape, speed and lifting capacity over any airship commissioned in this country, and as a whole the workmanship exhibited in her construction was exquisite. Unfortunately, under the conditions to which it was subjected, the hull structure did not prove durable, and to those conditions the failure is attributed. Under different circumstances it may be hoped that the second ship, when completed, will prove more fortunate.

In length R 31 was 615 feet, with a diameter of 66 feet, and the capacity was 1 1/2 million cubic feet.

In shape the hull was similar to the later types of Zeppelin, having a rounded bow and a long, tapering stern. The longitudinal and transverse frames were composed of girders built up of three-ply wood, the whole structure being braced in the usual manner with wire bracings. It had been found in practice with rigid airships that, if for any reason one gasbag becomes much less inflated than those adjacent to it, there is considerable pressure having the effect of forcing the radial wires of the transverse frames towards the empty bag. The tension resulting in these wires may produce very serious compressive strain in the members of the transverse frames, and to counteract this action an axial wire is led along the axis of the ship and secured to the centre point of the radial wiring. This method, now current practice in rigid airship construction, was introduced for the first time in this ship.

As will be seen from the photograph, the control and navigating compartment of the ship is contained in the hull, the cars in each case being merely small engine rooms. These small cars were beautifully made of wood of a shape to afford the least resistance to the air, and in number were five, each housing a single 250 horse-power Rolls Royce engine driving a single fixed propeller. Here we see another decided departure from our previous methods of rigid airship construction, in that for the first time swivelling propellers were abandoned. R 31 when completed carried out her trials, and it was evident that she was much faster than previous ships. The trials were on the whole satisfactory and, except for a few minor accidents to the hull framework and fins, nothing untoward occurred.

At a later date the whole ship was through fortuitous circumstances exposed to certain disadvantageous conditions which rendered her incapable of further use.

## RIGID AIRSHIP No. 33 CLASS

September 24th, 1916, is one of the most important days in the history of rigid airship design in this country; on this date the German Zeppelin airship L 33 was damaged by gunfire over London, and being hit in the after gasbags attempted to return to Germany. Owing to lack of buoyancy she was forced to land at Little Wigborough, in Essex, where the crew, having set fire to the ship, gave themselves up. Although practically the entire fabric of the ship was destroyed, the hull structure most fortunately remained to all intents and purposes intact, and was of inestimable value to the design staff of the Admiralty, who measured up the whole ship and made working drawings of every part available.

During this year other German rigid airships had been brought down, namely L 15, which was destroyed at the mouth of the Thames in April, but which was of an old type, and from which little useful information was obtained; and also the Army airship L.Z. 85, which was destroyed at Salonica in the month of May. A Schutte-Lanz airship was also brought down at Cuffley, on September 2nd, and afforded certain valuable details.

All these ships were, however, becoming out of date; but L 33 was of the latest design, familiarly called the super-Zeppelin, and had only been completed about six weeks before she encountered disaster.

In view of the fact that the rigid airships building in this country at this date, with the exception of the wooden Schutte-Lanz ships were all based on pre-war designs of Zeppelin airships, it can be readily understood that this latest capture revolutionized all previous ideas, and to a greater extent than might be imagined, owing to the immense advance, both in design and construction, which had taken place in Germany since 1914.

All possible information having been obtained, both from the wreck of the airship itself and from interrogation of the captured crew, approval was obtained, in November of the same year, for two ships of the L 33 design to be built; and in January, 1917, this number was increased to five.

It was intended originally that these ships should be an exact facsimile of L 33; but owing to the length of time occupied in construction later information was obtained before they were completed, both from ships of a more modern design, which were subsequently brought down, and also from other sources. Acting on this information, various improvements were embodied in R 33 and R 34, which were in a more advanced state; but in the case of the three other ships the size was increased, and the ships, when completed, will bear resemblance to a later type altogether.

As a comment on the slowness of construction before mentioned, the fact that while we in this country were building two ships on two slips, Germany had constructed no fewer than thirty on four slips, certainly affords considerable food for reflection.

The two airships of this class having only just reached a state of completion, a detailed description cannot be given without making public much information which must necessarily remain secret for the present. Various descriptions have, however, been given in the daily and weekly Press, but it is not intended in the present edition of this book to attempt to elaborate on anything which has not been already revealed through these channels.

It is regrettable that so much that would be of the utmost interest has to be omitted; but the particulars which follow will at any rate give sonic idea of the magnitude of the ship and show that it marks a decided departure from previous experiments and a great advance on any airship before constructed in Great Britain.

It is also a matter for regret that these two ships were not completed before the termination of hostilities, as their capabilities would appear to be sufficient to warrant the expectations which have been based on their practical utility as scouting agents for the Grand Fleet.

In all its main features the hull structure of R 33 and R 34 follows the design of the wrecked German Zeppelin airship L 33. The hull follows more nearly a true streamline shape than in the previous ships constructed of duralumin, in which a great proportion of the total length was parallel-sided. The Germans adopted this new shape from the Schutte-Lanz design and have not departed from this practice. This consists of a short parallel body with a long rounded bow and a long tapering stem culminating in a point. The overall length of the ship is 643 feet with a diameter of 79 feet and an extreme height of 92 feet.

The type of girders in this class has been much altered from those in previous ships. The hull is fitted with an internal triangular keel throughout practically the entire length. This forms the main corridor of the ship, and is fitted with a footway down the centre for its entire length. It contains water ballast and petrol tanks, bomb stowage and crew accommodation and the various control wires, petrol pipes and electric leads are carried along the lower part.

Throughout this internal corridor runs a bridge girder, from which the petrol and water ballast tanks are supported. These tanks are so arranged that they can be dropped clear of the ship.

Amidships is the cabin space with sufficient room for a crew of twenty-five. Hammocks can be slung from the bridge girder before mentioned.

In accordance with the latest Zeppelin practice, monoplane rudders and elevators are fitted to the horizontal and vertical fins.

The ship is supported in the air by nineteen gasbags which give a total capacity of approximately two million cubic feet of gas. The gross lift works out at approximately 59 1/2 tons, of which the total fixed weight is 33 tons, giving a disposable lift of 26 1/2 tons.

The arrangement of cars is as follows: At the forward end the control car is slung, which contains all navigating instruments and the various controls. Adjoining this is the wireless cabin, which is also fitted for wireless telephony. Immediately aft of this is the forward power car containing one engine, which gives the appearance that the whole is one large car.

Amidships are two wing cars each containing a single engine. These are small and just accommodate the engine with sufficient room for mechanics to attend to them. Further aft is another larger car which contains an auxiliary control position and two engines.

It will thus be seen that five engines are installed in the ship; these are all of the same type and horse-power, namely, 250 horse-power Sunbeam. R 33 was constructed by Messrs. Armstrong Whitworth Ltd., while her sister ship R 34 was built by Messrs. Beardmore on the Clyde.

In the spring of 1918, R 33 and R 34 carried out several flight trials, and though various difficulties were encountered both with the engines and also with the elevator and rudder controls, it was evident that, with these defects remedied, each of these ships would prove to be singularly reliable.

On one of these trials made by R 34, exceedingly bad weather was encountered, and the airship passed through several blinding snowstorms; nevertheless the proposed flight of some seventeen hours was completed, and though at times progress was practically nil owing to the extreme force of the wind, the station was reached in safety and the ship landed without any contretemps. This trial run having been accomplished in weather such as would never have been chosen in the earlier days of rigid trial flights, those connected with the airship felt that their confidence in the vessel's capabilities was by no means exaggerated.

The lift of the ship warranted a greater supply of petrol being carried than there was accommodation for, and the engines by now had been "tuned up" to a high standard of efficiency. Accordingly it was considered that the ship possessed the necessary qualifications for a transatlantic flight. It was, moreover, the opinion of the leading officers of the airship service that such an enterprise would be of inestimable value to the airship itself, as demonstrating its utility in the future for commercial purposes.

Efforts were made to obtain permission for the flight to be attempted, and although at first the naval authorities were disinclined to risk such a valuable ship on what appeared to be an adventure of doubtful outcome, eventually all opposition was overcome and it was agreed that for the purposes of this voyage the ship was to be taken over by the Air Ministry from the Admiralty.

Work was started immediately to fit out the ship for a journey of this description. Extra petrol tanks were disposed in the hull structure to enable a greater supply of fuel to be carried, a new and improved type of outer cover was fitted, and by May 29th, R 34 was completed to the satisfaction of the Admiralty and was accepted. On the evening of the same day she left for her station, East Fortune, on the Firth of Forth. This short passage from the Clyde to the Forth was not devoid of incident, as soon after leaving the ground a low-lying fog enveloped the whole country and it was found impossible to land with any degree of safety. It having been resolved not to land until the fog lifted, the airship cruised about the northeast coast of England and even came as far south as York. Returning to Scotland, she found the fog had cleared, and was landed safely, having been in the air for 21 hours.

The original intention was that the Atlantic flight should be made at the beginning of June, but the apparent unwillingness of the

Germans to sign the Peace Treaty caused the Admiralty to retain the ship for a time and commission her on a war footing. During this period she went for an extended cruise over Denmark, along the north coast of Germany and over the Baltic. This flight was accomplished in 56 hours, during which extremely bad weather conditions were experienced at times. On its conclusion captain and crew of the ship expressed their opinion that the crossing of the Atlantic was with ordinary luck a moral certainty. Peace having been signed, the ship was overhauled once more and made ready for the flight, and the day selected some three weeks before was July 2nd.

A selected party of air-service ratings, together with two officers, were sent over to America to make all the necessary arrangements, and the American authorities afforded every conceivable facility to render the flight successful.

As there is no shed in America capable of housing a big rigid, there was no alternative but to moor her out in the open, replenish supplies of gas and fuel and make the return journey as quickly as possible.

On July 2nd, at 2.38 a.m. (British summer time), R 34 left the ground at East Fortune, carrying a total number of 30 persons. The route followed was a somewhat northerly one, the north coast of Ireland being skirted and a more or less direct course was kept to Newfoundland. From thence the south-east coast of Nova Scotia was followed and the mainland was picked up near Cape Cod.

From Cape Cod the airship proceeded to Mineola, the landing place on Long Island. All went well until Newfoundland was reached. Over this island fog was encountered, and later electrical storms became a disturbing element when over Nova Scotia and the Bay of Fundy. The course had to be altered to avoid these storms, and owing to this the petrol began to run short. No anxiety was occasioned until on Saturday, July 5th, a wireless signal was sent at 3.59 p.m. asking for assistance, and destroyers were dispatched immediately to the scene. Later messages were received indicating that the position was very acute, as head winds were being encountered and petrol was running short. The airship, however, struggled on, and though at one time the possibility of landing at Montauk, at the northern end of Long Island, was considered, she managed after a night of considerable anxiety to reach Mineola and land there in safety on July 6th at 9.55 a.m. (British summer time). The total duration of the outward voyage was 108 hours 12 minutes, and during this time some 3,136 sea miles were covered. R 34 remained at Mineola until midnight of July 9th according to American time. During the four days in which she was moored out variable weather was experienced, and in a gale of wind the mooring point was torn out, but fortunately, another trail rope was dropped and made fast, and the airship did not break away.

It was intended that the return should be delayed until daylight, in order that spectators in New York should obtain a good view of the airship, but an approaching storm was reported and the preparations were advanced for her immediate departure. During the last half-hour great difficulty was experienced in holding the ship while gassing was completed.

At 5.57 a.m. (British summer time) R 34 set out on her return voyage, steering for New York, to fly over the city before heading out into the Atlantic. She was picked up by the searchlights and was distinctly visible to an enormous concourse of spectators. During the early part of the flight a strong following wind was of great assistance, and for a short period an air speed of 83 miles per hour was attained. On the morning of July 11th the foremost of the two engines in the after car broke down and was found to be beyond repair. The remainder of the voyage was accomplished without further incident. On July 12th at noon, a signal was sent telling R 34 to proceed to the airship station at Pulham in Norfolk as the weather was unfavourable for landing in Scotland. On the same day at 8.25 p.m., land was first sighted and the coast line was crossed near Clifden, county Galway, at 9 p.m. On the following morning, July 13th, at 7.57 a.m. (British summer time), the long voyage was completed and R 34 was safely housed in the shed, having been in the air 75 hours 3 minutes.

Thus a most remarkable undertaking was brought to a successful conclusion. The weather experienced was by no means abnormally good. This was not an opportunity waited for for weeks and then hurriedly snatched, but on the preordained date the flight was commenced. The airship enthusiast had always declared that the crossing of the Atlantic presented no insuperable difficulty, and when the moment arrived the sceptics found that he was correct. We may therefore assume that this flight is a very important landmark in the history of aerial transport, and has demonstrated that the airship is to be the medium for long-distance travel. We may rest assured that such flights, although creating universal wonder to-day, will of a surety be accepted as everyday occurrences before the world is many years older.

# CHAPTER VIII

## THE WORK OF THE AIRSHIP IN THE WORLD WAR

The outbreak of war found us, as we have seen, practically without airships of any military value. For this unfortunate circumstance there were many contributory causes. The development of aeronautics generally in this country was behind that of the Continent, and the airship had suffered to a greater extent than either the seaplane or the aeroplane. Our attitude in fact towards the air had not altered so very greatly from that of the man who remarked, on reading in his paper that some pioneer of aviation had met with destruction, "If we had been meant to fly, God would have given us wings." Absurd as this sounds nowadays, it was the opinion of most people in this country, with the exception of a few enthusiasts, until only a few years before we were plunged into war.

The year 1909 saw the vindication of the enthusiasts, for in this summer Bleriot crossed the Channel in an aeroplane, and the first passenger-carrying Zeppelin airship was completed. Those who had previously scoffed came to the conclusion that flying was not only possible but an accomplished fact, and the next two years with their great aerial cross-country circuits revealed the vast potentialities of aircraft in assisting in military operations. We, therefore, began to study aeronautics as the science of the future, and aircraft as an adjunct to the sea and land forces of the empire.

The airship, unfortunately, suffered for many reasons from the lack of encouragement afforded generally to the development of aeronautics. The airship undoubtedly is expensive, and one airship of size costs more to build than many aeroplanes. In addition, everything connected with the airship is a source of considerable outlay. The shed to house an airship is a most costly undertaking, and takes time and an expenditure of material to erect, and bears no comparison with the cheap hangar which can be run up in a moment to accommodate the aeroplane. The gas to lift the airship is by no means a cheap commodity. If it is to be made on the station where the airship is based, it necessitates the provision of an expensive and elaborate plant. If, on the other hand, it is to be manufactured at a factory, the question of transport comes in, which is a further source of expense with costly hydrogen tubes for its conveyance.

Another drawback is the large tract of ground required for an aerodrome, and the big airship needs a large number of highly-trained personnel to handle it.

A further point always, raised when the policy of developing the airship was mooted is its vulnerability. It cannot be denied that it presents a large target to artillery or to the aeroplane attacking it, and owing to the highly inflammable nature of hydrogen when mixed with air there can be no escape if the gas containers are pierced by incendiary bullets or shells.

Another contributing factor to the slow development of the airship was the lack of private enterprise. Rivalry existed between private firms for aeroplane contracts which consequently produced improvements in design; airships could not be produced in this way owing to the high initial cost, and if the resulting ships ended in failure, as many were bound to do, there would be no return for a large outlay of capital. The only way by which private firms could be encouraged to embark on airship building was by subsidies from the Government, and at this time the prevalent idea of the doubtful value of the airship was too strong for money to be voted for this purpose.

To strengthen this argument no demand had either been made from those in command of the Fleet or from commanders of our Armies for airships to act as auxiliaries to our forces.

The disasters experienced by all early airships and most particularly by the Zeppelins were always seized upon by those who desired to convince the country what unstable craft they were, and however safe in the air they might be were always liable to be wrecked when landing in anything but fine weather. Those who might have sunk their money in airship building thereupon patted themselves upon the back and rejoiced that they had been so far-seeing as to avoid being engaged upon such a profitless industry.

Finally, all in authority were agreed to adopt the policy of letting other countries buy their experience and to profit from it at a later date. Had the war been postponed for another twenty years all might have been well, and we should have reaped the benefit, but most calamitously for ourselves it arrived when we were utterly unprepared, and having, as we repeat, only three airships of any military value.

With these three ships, Astra-Torres (No. 3), Parseval (No. 4) and Beta, the Navy did all that was possible. At the very outbreak of war scouting trips were made out into the North Sea beyond the mouth of the Thames by the Astra and Parseval, and both these ships patrolled the Channel during the passage of the Expeditionary Force.

The Astra was also employed off the Belgian coast to assist the naval landing party at Ostend, and together with the Parseval assisted in patrolling the Channel during the first winter of the war.

The Beta was also sent over to Dunkirk to assist in spotting for artillery fire and locating German batteries on the Belgian coast. Our airships were also employed for aerial inspection of London and other large towns by night to examine the effects of lighting restrictions and obtain information for our anti-aircraft batteries.

With the single exception of the S.S. ship, which carried out certain manoeuvres in France in the summer of 1916, our airships were confined to operations over the sea; but if we had possessed ships of greater reliability in the early days of the war, it is conceivable that they would have been of value for certain purposes to the Army. The Germans employed their Zeppelins at the bombardment of Antwerp, Warsaw, Nancy and Libau, and their raids on England are too well remembered to need description. The French also used airships for the observation of troops mobilizing and for the destruction of railway depots. The Italians relied entirely at the beginning of the war on airships, constructed to fly at great heights, for the bombing of Austrian troops and territory, and met with a considerable measure of success.

When it was decided, early in 1915, to develop the airship for anti-submarine work difficulties which appeared almost insuperable were encountered at first. To begin with, there were practically no firms in the country capable of airship production. The construction of envelopes was a great problem; as rubber-proofed fabric had been found by experiment to yield the best results for the holding of gas, various waterproofing firms were invited to make envelopes, and by whole-hearted efforts and untiring industry they at last provided very excellent samples. Fins, rudder planes, and cars were also entrusted to firms which had had no previous experience of this class of work, and it is rather curious to reflect that envelopes were produced by the makers of mackintoshes and that cars and planes were constructed by a shop-window furnisher. This was a sure sign that all classes of the community were pulling together for the good of the common cause.

Among other difficulties was the shortage of hydrogen tubes, plants, and the silicol for making gas.

Sufficient sheds and aerodromes were also lacking, and the airships themselves were completed more quickly than the sheds which were to house them.

The lack of airship personnel to meet the expansion of the service presented a further obstacle. To overcome this the system of direct entry into the R.N.A.S. was instituted, which enabled pilots to be enrolled from civil life in addition to the midshipmen who were drafted from the Fleet. The majority of the ratings were recruited from civil life and given instruction in rigging and aero-engines as quickly as possible, while technical officers were nearly all civilians and granted commissions in the R.N.V.R.

A tremendous drawback was the absence of rigid airships and the lack of duralumin with which to construct them.

Few men were also experienced in airship work at this time, and there was no central airship training establishment as was afterwards instituted. Pilots were instructed as occasion permitted at the various patrol stations, having passed a balloon course and undergone a rudimentary training at various places.

To conclude, the greatest of all difficulties was the shortage of money voted for airship development, and this was a disadvantage under which airships laboured even until the conclusion of hostilities.

We have seen previously how the other difficulties were surmounted and how our airships were evolved, type by type, and the measure of success which attended them. It is interesting to recall that five years ago we only possessed three ships capable of flying, and that during the war we built upwards of two hundred, of which no fewer than 103 were actually in commission on the date of the signing of the Armistice.

The work carried out by our airships during the war falls under three main headings:

1. Operations with the fleet or with various units.

2. Anti-submarine patrol and searching for mines.

3. Escort of shipping and examination duties.

With regard to the first heading it is only permissible at present to say very little; certain manoeuvres were carried out in connection with the fleet, but the slow development of our rigid airships prohibited anything on a large scale being attempted. The Germans, on the other hand, made the fullest use of their Zeppelins for scouting purposes with the high seas fleet. Responsible people were guilty of a grave mistake when speaking in public in denouncing the Zeppelin as a useless monster every time one was destroyed in a raid on this country. The main function of the Zeppelin airship was to act as an aerial scout, and it carried out these duties with the utmost efficiency during the war. It is acknowledged that the German fleet owed its escape after the Battle of Jutland to the information received from their airships, while again the Zeppelin was instrumental in effecting the escape of the flotilla which bombarded Scarborough in 1916.

Very probably, also, the large airship was responsible for the success which attended the U boats during their attack on the cruisers Nottingham and Falmouth, and also at the Hogue disaster.

Various experiments were carried out in towing airships by cruisers, in refuelling while in tow and changing crews, all of which would have borne good fruit had the war lasted longer.

An exceedingly interesting experiment was carried out during the closing stages of the war by an airship of the S.S. Zero type. At this period the German submarines were gradually extending their operations at a greater distance from our coasts, and the authorities became concerned at the prospect that the small type of airship would not possess sufficient endurance to carry out patrol over these increased distances. The possibility was considered of carrying a small airship on board a ship which should carry out patrol and return to the ship for refuelling purposes, to replenish gas, and change her crew. To test the feasibility of this idea S.S. Z 57 carried out landing experiments on the deck of H.M.S. Furious, which had been adapted as an aeroplane carrier. S.S. Z 57 came over the deck and dropped her trail rope, which was passed through a block secured to the deck, and was hauled down without difficulty. These experiments were continued while the ship was under weigh and were highly successful. No great difficulty was encountered in making fast the trail rope, and the airship proved quite easy to handle. The car was also lowered into the hangar below the upper deck, the envelope only remaining on the upper level, and everything worked smoothly. If the war had continued there is no doubt that some attempt would have been made to test the practical efficiency of the problem.

Anti-submarine patrol was the chief work of the airship during the war, and, like everything else, underwent most striking changes. Submarine hunting probably had more clever brains concentrated upon it than anything else in the war, and the part allotted to the airship in conjunction with the hunting flotillas of surface craft was carefully thought out.

In the case of a suspected submarine in a certain spot, all surface and air craft were concentrated by means of wireless signals at the appointed rendezvous. It is in operations of this kind that the airship is so superior to the seaplane or aeroplane, as she can hover over a fixed point for an indefinite period with engines shut off. If the submarine was located from the air, signals were given and depth charges dropped in the position pointed out. Incidents of this kind were of frequent occurrence, and in them the value of the airship was fully recognized.

The most monotonous and arduous of the airship's duties was the routine patrol. The ship would leave her shed before dawn and be at the appointed place many miles away from land. She then would carry out patrol, closely scanning the sea all round, and investigating any suspicious object. For hours this might last with nothing seen, and then in the gathering darkness the ship would make her way home often against a rising wind, and in the winter through hail and snow. Bombs were always carried, and on many occasions direct hits were observed on enemy submarines. A sharp look-out was always kept for mines, and many were destroyed, either by gunfire from the airship herself or through the agency of patrol boats in the vicinity. This was the chief work of the S.S. ships, and was brought to a high pitch of perfection by the S.S. Zero. These ships proved so handy that they could circle round an object without ever losing sight of it, and yet could be taken in and out of sheds in weather too bad to handle bigger ships.

The hunting of the submarine has been likened to big-game hunting, and certainly no one ever set out to destroy a bigger quarry. It needs the same amount of patience and the same vigilance. Days may pass without the opportunity, and that will only be a fleeting one: the psychological moment must be seized and it will not brook a moment's delay. The eye must be trained to pick up the minutest detail, and must be capable of doing this for hour after hour. For those on submarine patrol in a small ship there is not one second's rest. As is well known, the submarine campaign reached its climax in April, 1917. In that month British and Allied shipping sustained its greatest losses. The value of the airship in combating this menace was now fully recognized, and with the big building programme of Zero airships approved, the housing accommodation again reached an acute stage.

Shortage of steel and timber for shed building, and the lack of labour to erect these materials had they been available, rendered other methods necessary. It was resolved to try the experiment of mooring airships in clearings cut into belts of trees or small woods.

A suitable site was selected and the trees were felled by service labour. The ships were then taken into the gaps thus formed and were moored by steel wires to the adjacent trees. Screens of brushwood were then built up between the trees, and the whole scheme proved so successful that even in winter, when the trees were stripped of their foliage, airships rode out gales of over 60 miles per hour. The personnel were housed either in tents or billeted in cottages or houses in the neighbourhood, and gas was supplied in tubes as in the earlier days of the stations before the gas plants had been erected.

This method having succeeded beyond the most sanguine expectations, every station had one or more of these sub-stations based on it, the airships allocated to them making a periodical visit to the parent station for overhaul as required. Engineering repairs were effected by workshop lorries, provided that extensive work was not required.

In this way a large fleet of small airships was maintained around our coasts, leaving the bigger types of ships on the parent stations, and the operations were enabled to be considerably extended. Of course, certain ships were wrecked when gales of unprecedented violence sprung up; but the output of envelopes, planes and cars was by this time so good that a ship could be replaced at a few hours' notice, and the cost compared with building of additional sheds was so small as to be negligible.

From the month of April, 1917, the convoy system was introduced, by which all ships on entering the danger zones were collected at an appointed rendezvous and escorted by destroyers and patrolboats. The airship was singularly suitable to assist in these duties. Owing to her power of reducing her speed to whatever was required, she could keep her station ahead or abeam of the convoy as was necessary, and from her altitude was able to exercise an outlook for a far greater distance than was possible from the bridge of a destroyer. She could also sweep the surface ahead of the approaching convoy, and warn it by wireless or by flash-lamp of the presence of submarines or mines. By these timely warnings many vessels were saved. Owing to the position of the stations it was possible for a convoy to be met by airships west of the Scilly Isles and escorted by the airships of the succeeding stations right up the Channel. In a similar manner, the main shipping routes on the east coast and also in the Irish Sea were under constant observation. The mail steamers between England and Ireland and transports between England and France were always escorted whenever flying conditions were possible. For escort duties involving long hours of flying, the Coastal and C Star types were peculiarly suitable, and at a later date the North Sea, which could accompany a convoy for the length of Scotland. Airships have often proved of value in summoning help to torpedoed vessels, and on occasions survivors in open boats have been rescued through the agency of patrolling airships. Examination duties are reckoned among the many obligations of the airship. Suspicious-looking vessels were always carefully scrutinized, and if unable to give a satisfactory answer to signals made, were reported to vessels of the auxiliary patrol for closer examination. Isolated fishing vessels always were kept under close observation, for one of the many ruses of the submarine was to adopt the disguise of a harmless fishing boat with masts and sails.

The large transports, conveying American troops who passed through England on their way to France, were always provided with escorting airships whenever possible, and their officers have extolled their merits in most laudatory terms.

Our rigid airships also contributed their share in convoy work, although their appearance as active units was delayed owing to slowness in construction.

A disturbing feature to the advocate of the large airship, has been the destruction of raiding Zeppelins by heavier-than-air machines, and the Jeremiahs have not lost this opportunity of declaring that for war purposes the huge rigid is now useless and will always be at the complete mercy of the fast scouting aeroplane. There is never any obstacle in this world that cannot be surmounted by some means or other. On the one hand there is helium, a non-inflammable gas which would render airships almost immune to such attacks. On the other hand, one opinion of thought is that the rigid airship in the future will proceed to sea escorted by a squadron of scouting aeroplanes for its defence, in the same way that the capital ship is escorted at sea by destroyers and torpedo boats. This latter idea has been even further developed by those who look into the future, and have conceived the possibility of a gigantic airship carrying its own aeroplanes for its protection.

To test the possibility of this innovation, a small aeroplane was attached to one of our rigid airships beneath the keel. Attachments were made to the top of the wings and were carried to the main framework of the hull. The release gear was tested on the ground to preclude the possibility of any accident, and on the day appointed the airship was got ready for flight. While the airship was flying, the pilot of the aeroplane was in his position with his engine just ticking over. The bows of the airship were then inclined upwards and the release gear was put into operation. The pilot afterwards said that he had no notion that anything had been done until he noticed that the airship was some considerable height above him. The machine made a circuit of the aerodrome and landed in perfect safety, while no trouble was experienced in any way in the airship. Whether this satisfactory experiment will have any practical outcome the future alone can say, but this achievement would have been considered, beyond all the possibilities of attainment only a few years ago.

Since the Armistice several notable endurance flights were accomplished by ships of the North Sea class, several voyages being made to the coast of Norway, and quite recently a trip was carried out all round the North Sea.

The weather has ceased to be the deterrent of the early days. Many will no doubt remember seeing the North Sea airship over London on a day of squalls and snow showers, and R 34 encountered heavy snow storms on the occasion of one of her flight trials, which goes to prove that the airship is scarcely the fair-weather aircraft as maintained by her opponents.

Throughout the war our airships flew for approximately 89,000 hours and covered a distance of upwards of two and a quarter million miles. The Germans attempted to win the war by the wholesale sinking of our merchant shipping, bringing supplies and food to these islands, and by torpedoing our transports and ships carrying guns and munitions of war. They were, perhaps, nearer to success than we thought at the time, but we were saved by the defeat of the submarine. In the victory won over the underseas craft the airship certainly played a prominent part and we, who never suffered the pinch of hunger, should gratefully remember those who never lost heart, but in spite of all difficulties and discouragement, designed, built, maintained and flew our fleet of airships.

# CHAPTER IX

## THE FUTURE OF AIRSHIPS

With the signing of the Armistice on November 11th, 1918, the airship's work in the war was practically completed and peace reigned on the stations which for so many months had been centres of feverish activity. The enemy submarines were withdrawn from our shipping routes and merchant ships could traverse the sea in safety except for the occasional danger of drifting mines. "What is to be the future of the airship?" is the question which is agitating the minds of innumerable people at the present moment.

During the war we have built the largest fleet of airships in the world, in non-rigids we have reached a stage in design which is unsurpassed by any country, and in rigid airships we are second only to the Germans, who have declared that, with the signing of the peace terms, their aircraft industry will be destroyed. Such is our position at the present moment, a position almost incredible if we look back to the closing days of the year 1914. Are we now to allow ourselves to drift gradually back to our old policy of supineness and negligence as existed before the war? Surely such a thought is inconceivable; as we have organized our airship production for the purposes of war, so shall we have to redouble our efforts for its development in peace, if we intend to maintain our supremacy in the air.

Unless all war is from henceforth to cease, a most improbable supposition when the violence of human nature is considered, aircraft will be in the future almost the most important arm. Owing to its speed, there will not be that period of waiting for the concentration and marching of the armies of the past, but the nation resolved on war will be able to strike its blow, and that a very powerful and terrible one, within a few hours of the rupture of negotiations. Every nation to be prepared to counter such a blow must be possessed of adequate resources, and unless the enormous expense is incurred of maintaining in peace a huge establishment of aircraft and personnel, other methods must be adopted of possessing both of these available for war while employed in peace for other purposes.

From the war two new methods of transportation have emerged--the aeroplane and the airship. To the business man neither of these is at the present juncture likely to commend itself on the basis of cost per ton mile. When, however, it is considered that the aeroplane is faster than the express train and the airship's speed is double that of the fastest merchant ship, it will be appreciated that for certain commercial purposes both these mediums for transport have their possibilities. The future may prove that in the time to come both the airship and the aeroplane will become self-supporting, but for the present, if assisted by the Government, a fair return may be given for the capital laid out, and a large fleet of aircraft together with the necessary personnel will always be available for military purposes should the emergency arise. The present war has shown that the merchant service provided a valuable addition both of highly-trained personnel and of vessels readily adapted for war purposes, and it appears that a similar organization can be effected to reinforce our aerial navies in future times of danger.

In discussions relative to the commercial possibilities of aircraft, a heated controversy always rages between advocates of the airship and those of the heavier-than-air machine, but into this it is not proposed to plunge the reader of this volume. The aeroplane is eminently adapted for certain purposes, and the greatest bigot in favour of the airship can hardly dispute the claims of this machine to remain predominant for short-distance travel, where high speed is essential and the load to be carried is light. For long distance voyages over the oceans or broken or unpopulated country, where large loads are to be carried, the airship should be found to be the more suitable.

The demand for airships for commercial purposes falls under three main headings, which will be considered in some detail. It will be shown to what extent the present types will fill this demand, and how they can be developed in the future to render the proposed undertakings successful.

1. Pleasure.

2. A quick and safe means of transport for passengers.

3. A quick commercial service for delivering goods of reasonable weight from one country to another.

1. Pleasure.--In the past, men have kept mechanically-driven means of transport such as yachts, motor cars, and motor boats for their amusement, and to a limited extent have taken recreation in the air by means of balloons. For short cruises about this country and round the coast a small airship, somewhat similar to the S.S. Zero, would be an ideal craft. In cost it would be considerably less than a small yacht, and as it would only be required in the summer months, it would be inflated and moored out in the open in a park or grounds and the expense of providing a shed need not be incurred. For longer distances, a ship of 150,000 cubic feet capacity, with a covered-in car and driven by two engines, would have an endurance of 25 hours at a cruising speed of 45 miles per hour. With such a ship voyages could easily be made from the south coast to the Riviera or Spain, and mooring out would still be possible under the lee of a small wood or to a buoy on the water.

Possibilities also exist for an enterprising firm to start a series of short pleasure trips at various fashionable seaside resorts, and until the novelty had worn off the demand for such excursions will probably be far in excess of the supply.

2. Passenger transport.--In the re-organization of the world after this devastating war the business man's time will be of even more value than it was before. This country is largely bound up with the United States of America in business interests which necessitate continual visits between the two countries. The time occupied by steamer in completing this journey is at present about five days. If this time can be cut down to two and a half days, no doubt a large number of passengers will be only too anxious to avail themselves of this means of travel, providing that it will be accomplished in reasonable safety and comfort. The requirements for this purpose are an aerial liner capable of carrying a hundred passengers with a certain quantity of luggage and sufficient provisions for a voyage which may be extended over the specified time owing to weather conditions. The transatlantic service if successful could then be extended until regular passenger routes are established encircling the globe.

3. Quick commercial service for certain types of goods.--Certain mails and parcels are largely enhanced in value by the rapidity of

transport, and here, as in the passenger service outlined above, the airship offers undoubted facilities. As we have said before, it is mainly over long distances that the airship will score, and for long distances on the amount carried the success of the enterprise will be secured. For this purpose the rigid airship will be essential. There are certain instances in which the non-rigid may possibly be profitably utilized, and one such is suggested by a mail service between this country and Scandinavia. A service is feasible between Newcastle and Norway by airships of a capacity of the S.S. Twin type. These ships would carry 700 lb. of mails each trip at about 4d. per ounce, which would reduce the time of delivering letters from about two and a half to three days to twenty-four hours.

A commercial airship company is regarded in this country as a new and highly hazardous undertaking, and it seems to be somewhat overlooked that it is not quite the novel idea so many people imagine. Before the war, in the years 1910 to 1914, the Deutsche Luftfahrt Actien Gesellschaft successfully ran a commercial Zeppelin service in which four airships were used, namely, Schwaben, Victoria Luise, Hansa and Sachsan. During this period over 17,000 passengers were carried a total distance of over 100,000 miles without incurring a single fatal accident. Numerous English people made trips in these airships, including Viscount Jellicoe, but the success of the company has apparently been forgotten.

We have endeavoured to show that the non-rigid airship has potentialities even for commercial purposes, but there is no doubt whatever that the future of the airship in the commercial world rests entirely with the rigid type, and the airships of this type moreover must be of infinitely greater capacity than those at present in existence, if a return is to be expected for the capital invested in them. General Sykes stated, in the paper which he read before the London Chamber of Commerce, "that for commercial purposes the airship is eminently adapted for long-distance journeys involving non-stop flights. It has this inherent advantage over the aeroplane, that while there appears to be a limit to the range of the aeroplane as at present constructed, there is practically no limit whatever to that of the airship, as this can be overcome by merely increasing the size. It thus appears that for such journeys as crossing the Atlantic, or crossing the Pacific from the west coast of America to Australia or Japan, the airship will be peculiarly suitable."

He also remarked that, "it having been conceded that the scope of the airship is long-distance travel, the only type which need be considered for this purpose is the rigid. The rigid airship is still in an embryonic state, but sufficient has already been accomplished in this country, and more particularly in Germany, to show that with increased capacity there is no reason why, within a few years' time, airships should not be built capable of completing the circuit of the globe and of conveying sufficient passengers and merchandise to render such an undertaking a paying proposition."

The report of the Civil Aerial Transport Committee also states that, "airships are the most suitable aircraft for the carrying of passengers where safety, comfort and reliability are essential."

When we consider the rapid development of the rigid airship since 1914, it should not be insuperable to construct an airship with the capabilities suggested by General Sykes. In 1914, the average endurance of the Zeppelin at cruising speed was under one day and the maximum full speed about 50 miles per hour. In 1918, the German L 70, which is of 2,195,000 cubic feet capacity, the endurance at 45 miles per hour has risen to 7.4 days and the maximum full speed to 77 miles per hour. The "ceiling" has correspondingly increased from 6,000 feet to 23,000 feet.

The British R 38 class, at present building, with a capacity of approximately 2 3/4 million cubic feet has an estimated endurance at 45 miles per hour of 211 hours or 8.8 days, which is 34 hours greater than the German L 70 class. It is evident that for a ship of this calibre the crossing of the Atlantic will possess no difficulty, and as an instance of what has already been accomplished in the way of a long-distance flight the exploit of a Zeppelin airship based in Bulgaria during the war is sufficiently remarkable. This airship in the autumn of 1917 left the station at Jamboli to carry twelve tons of ammunition for the relief of a force operating in German East Africa. Having crossed the Mediterranean, she proceeded up the course of the Nile until she had reached the upper waters of this river. Information was then received by wireless of the surrender of the force, and that its commander, Von Lettow, was a fugitive in the bush. She thereupon set out for home and reached her station in safety, having been in the air 96 hours, or four days, without landing.

It is therefore patent that in R 33 and R 34 we possess two airships which can cross to America to-morrow as far as actual distance is concerned, but various other conditions are necessary before such voyages can be undertaken with any prospects of commercial success.

The distance between England and America must be roughly taken as 3,000 miles. It is not reasonable for airship stations to be situated either in the inaccessible extreme west of Ireland or among the prevailing fogs of Newfoundland.

Weather conditions must also be taken into account; head winds may prevail, rendering the forward speed of the ship to be small even with the engines running full out. In calculations it is considered that the following assumptions should be made:

1. At least 75 per cent additional petrol to be carried as would be necessary for the passage in calm air, should unfavourable weather conditions be met. This amount could be reduced to 50 per cent in future airships with a speed of upwards of 80 miles per hour.
2. About a quarter of the total discharge able lift of the ship should be in the form of merchandise or passengers to render the project a reasonable commercial proposition.

We will consider the commercial loads that can be carried by the German airship L 70 and our airships R 33 and R 38 under the conditions given above. Two speeds will be taken for the purposes of this comparison: normal full speed, or about 60 miles per hour, and cruising speed of 45 miles per hour.

> L 70.--At 60 miles per hour a distance of 3,000 miles will be accomplished in 50 hours. Fuel consumption about 13 tons + 9.75 tons (additional for safety) = 22.75 tons. Available lift for fuel and freight = 27.8 tons. Fuel carried = 22.75 " ------------ Balance for freight = 5 " about. ------------ At 45 miles per hour, distance will be accomplished in 66.6 hours. Fuel consumption about 10 tons + 7.5 tons additional = 17.5 tons. Available lift = 27.8 tons Fuel carried = 17.5 " ------------ Balance for

freight = 10 " about. ------------ R. 33.--At 60 miles per hour. Fuel consumption 14.25 tons + 10.68 tons additional = 24.93 tons. Lift available for fuel and freight = 21.5 tons. Fuel carried = 24.93 " ------------ Minus balance = 3. 43 " ------------ At 45 miles per hour. Fuel consumption 9.66 tons + 7.23 tons (17 tons approx.) Lift available for fuel and freight = 21.5 tons. Fuel carried = 17 " ------------ Balance for freight = 4.5 " ------------ R. 38.-Estimated only. At 60 miles per hour. Fuel consumption 20 tons + 15 tons additional = 35 tons. Lift available for fuel and freight = 42 tons. Fuel carried = 35 " ------------ Balance for freight = 7 " ------------ At 45 miles per hour. Fuel consumption 12 tons + 9 tons additional = 21 tons. Lift available for fuel and freight = 42 " Fuel carried = 21 " ------------ Balance for freight = 21 " ------------

It will thus be seen that at the faster speed small commercial loads can be carried by L 70 and R 38 and not at all in the case of R 33, that is assuming, of course, that the extra fuel is carried, of which 75 per cent of the total does not appear at all excessive in view of the weather continually experienced over the Atlantic.

At the cruising speed the loads naturally increase but still, in L 70, and more particularly in R 33, they are too small to be considered commercially. In R 38, however, the load that can be carried at cruising speed is sufficient to become a commercial proposition.

From this short statement it is evident that, by a comparatively small increase in volume, the lifting capacity of an airship is enormously increased, and it is in this subject that the airship possesses such undoubted advantage over the aeroplane. In the heavier-than-air machine there is no automatic improvement in efficiency resulting from greater dimensions. In the airship, however, this automatic improvement takes place in a very marked degree; for example, an airship of 10,000,000 cubic feet capacity has five times the lift of the present 2,000,000 cubic feet capacity rigid, but the length of the former is only 1.7 times greater, and therefore the weight of the structure only five times greater (1.7); that is, the weight of the structure is directly proportional to the total lift. Having seen that the total lift varies as the cube of the linear dimensions while air resistance, B.H.P.--other things being equal--vary as the square of the linear dimensions, it follows that the ratio "weight of machinery/total lift" decreases automatically.

In comparing the different methods of transport for efficiency, the resistance or thrust required is compared as a percentage of the total weight. The result obtained is known as the "co-efficient of tractive resistance." Experiments have shown that as the size of the airship increases, the co-efficient of tractive resistance decreases to a marked extent; with a proportionate increase in horse-power it is proportionally more economical for a 10,000,000 cubic feet capacity rigid to fly at 80 miles per hour than for a 2,000,000 cubic feet capacity to fly at 60 miles per hour.

As the ratio "weight structure/total lift" is in airships fairly constant, it follows that the ratio "disposable lift/total lift" increases with the dimensions.

It is therefore obvious that increased benefits are obtained by building airships of a larger size, and that the bigger the ship the greater will be its efficiency, providing, of course, that it is kept within such limits that it can be handled on the ground and manoeuvred in the air.

The proportion of the useful lift in a large rigid, that is the lift available for fuel, crew, passengers, and merchandise, is well over 50 per cent when compared with the gross lift. When the accompanying table is studied it will be seen that with airships of large capacity the available lift will be such that considerable weights of merchandise or passengers can be carried.

| Capacity in cubic feet | Gross Lift in tons | Length in feet | Diameter in feet |
| --- | --- | --- | --- |
| 2,000,000 | 60.7 | 643 | 79 |
| 3,000,000 | 91.1 | 736 | 90.4 |
| 4,000,000 | 121.4 | 810 | 99.5 |
| 5,000,000 | 151.8 | 872 | 107.2 |
| 6,000,000 | 182.2 | 927 | 113.9 |
| 7,000,000 | 212.5 | 976 | 119.9 |
| 8,000,000 | 242.8 | 1,021 | 125.5 |
| 9,000,000 | 273.3 | 1,061 | 130.4 |
| 10,000,000 | 303.6 | 1,100 | 135.1 |

In airships of their present capacity, in order to obtain the greatest amount of lift possible, lightness of construction has been of paramount importance. With this object in view duralumin has been used, and complicated girders built up to obtain strength without increase of weight. In a large ship with a considerable gain in lift, steel will probably be employed with a simpler form of girder work. In that way cheapness of construction will be effected together with increased rapidity of output, and in addition the strength of the whole structure should be increased.

The rigid airship of 10,000,000 cubic feet capacity will have a disposable lift of over 200 tons available for fuel, crew, passengers, and merchandise in such proportions as are desired. The endurance of such a ship at a cruising speed of 45 miles per hour will be in the neighbourhood of three weeks, with a maximum speed of 70 to 80 miles per hour, and a "ceiling" of some 30,000 feet can be reached. This will give a range of over 20,000 miles, or very nearly a complete circuit of the globe.

For commercial purposes the possibilities of such a craft are enormous, and the uses to which it could be put are manifestly of great importance. Urgent mails and passengers could be transported from England to America in under half the time at present taken by the steamship routes, and any city in the world could be reached from London in a fortnight.

In the event of war in the future, which may be waged with a nation situated at a greater distance from this country than was Germany, aircraft Of long endurance will be necessary both for scouting in conjunction with our fleets and convoy duties. The British Empire is widely scattered, and large tracts of ocean lie between the various colonies, all of which will require protection for the safeguarding of our merchant shipping. The provision of a force of these large airships will greatly add to the security of our out-lying dominions.

We have now reached a point where it is incumbent on us to face certain difficulties which beset the airship of large dimensions, and which are always magnified by its detractors. Firstly, there is the expense of sheds in which to house it; secondly, the large number of trained personnel to assist in landing and handling it when on the ground; thirdly, the risks attendant on the weather--for the airship is still considered the general public as a fair-weather craft; and fourthly, though this is principally in connection with its efficiency for military purposes, its vulnerability. We will deal with the four difficulties enumerated under these headings seriatim, and endeavour to show to what extent they may be surmounted if not entirely removed.

The solution of the first two problems may be summed up in two words: "mooring out"; on the success of this it is considered that the whole future of airships for commercial purposes rests. It will be essential that in every country which the airship visits on its voyages, one large central station is established for housing and repairs. The position of such a station is dependent on good weather conditions and the best railway facilities possible. In all respects this station will be comparable to a dry dock for surface vessels. The airship will be taken into the shed for overhaul of hull structure, renewing of gasbags or outer cover, and in short to undergo a periodical refit. The cost of a shed capable of housing two rigid airships, even at the present time, should not greatly exceed L500,000. This sum, though considerable, is but a small item compared with the cost of constructing docks to accommodate the Atlantic liner, and when once completed the cost of maintenance is small when weighed against the amount annually expended in dredging and making good the wear and tear of a dock.

Apart from these occasional visits to a shed, the airship, in the ordinary way at the end of a voyage, will pick up its moorings as does the big steamer, and land its passengers and cargo, at the same time replenishing its supplies of fuel, gas, provisions, etc., while minor repairs to the machinery can be carried out as she rides in the air.

A completely satisfactory solution of the mooring problem for the rigid airship has yet to reach its consummation. We saw in the previous chapter how, in the case of small non-rigids, they were sheltered in berths cut into woods or belts of trees, but for the rigid airship something more secure and less at the mercy of the elements is required.

At the present moment three systems of mooring are in an experimental stage: one, known as "the single-wire system," is now practically acknowledged to fall short of perfection; the second, "the three-wire system," and the third, "mooring to a mast," both have their champions, but it is probable that the last will be the one finally chosen, and when thoroughly tried out with its imperfections eliminated will satisfy the most exacting critics.

The single-wire system is at the same time the simplest and most obvious method which suggests itself, and means that the ship is secured by a wire cable attached to a suitable point in the ship and led to some fixed point on the ground. It has been found that an airship secured in this way requires constant attention, and that steering is always necessary to render her steady in the air. Considerable improvement is obtained if a dragging weight is added to the wire, as it tends to check to a considerable extent lateral motion of the bow of the ship.

The three-wire system is an adaptation and an improvement on the one previously mentioned. In this case the mooring point of the ship is attached to three long wire cables, which, when raised in the air, form a pyramid to the head of which the ship is attached. These wires are led to bollards which form in plan an equilateral triangle. The lift of the ship raises these wires off the ground, and if the ship is trimmed up by the bows she will be found to resist the action of the wind. A rigid airship moored out by this method remained in the open for a considerable time and rendered the future of this experiment most hopeful. It was resolved to continue these experiments by adding a subsidiary system of wires with running blocks, the whole wiring to form a polygon revolving round a fixed centre. The disadvantages of this method appear to be rather serious. It seems that great difficulty will always be found in picking up these moorings in a high wind, and though this also applies to the method with the mast, the initial obstacles do not appear to be so great. A powerful engine driving a winch will be necessary to raise these heavy wires from the ground, although of course the lift of the airship will assist in this. Secondly, the lowering of passengers and cargo will not be easy as the ship will not be rigidly secured. This, however, can probably be managed when experiments have reached a further stage, and at present the system may be said to present distinct possibilities.

The third system, that of mooring to a mast, possesses several features peculiar to itself, and not embraced by the other two, which should secure it prolonged investigations. The system is by no means new and has been tried from time to time for several years, but since the question of mooring in the open has been so ventilated and is now considered of such vital importance, these experiments have been continued, and in less spasmodic fashion than in the past. In a trial with a small non-rigid airship some months ago a signal success was achieved. The ship remained attached to a mast in open country with no protection whatsoever for six weeks in two of the worst months of the year. During this period two men only were required to look after the ship, which experienced gales in which the force of the wind rose to 52 miles per hour, and not the slightest damage was sustained.

Two or three methods of attaching the airship to the mast have been proposed, but the one which appears to be most practical is to attach the extreme bow point of the ship to some form of cap, in which the nose of the ship will fit, and will revolve round the top of the mast in accordance with the direction of the wind.

For large airships, employed as passenger and commerce carriers, we can imagine the mast advanced a stage further, and transformed into a tower with a revolving head. Incorporated in this tower will be a lift for passengers and luggage, pipes also will be led to the summit through which both gas and water can be pumped into the ship. With the airship rigidly held at the head of such a structure all the difficulties of changing crews, embarking and disembarking passengers, shipping and discharging cargo and also refuelling, vanish at once. Assuming the mooring problem solved with success, and we feel correct in this assumption, the first two of our difficulties automatically disappear. Sheds will only be necessary as repair depots and will not be extensively required, all intermediate stopping places being provided with masts and necessary arrangements for taking in gas, etc. At these intermediate stations the number of men employed will be comparatively speaking few. At the depots or repair stations the number must, of course, be considerably increased, but the provision of an enormous handling party will not be necessary. At present large numbers of men are only required to take a large airship in or out of a shed when the wind is blowing in a direction across the shed; when these conditions prevail the airship will, unless compelled by accident or other unforeseen circumstances, remain moored out in the open until the direction of the wind has changed.

Mechanical traction will also help effectually in handling airships on the ground, and the difficulty of taking them in and out of sheds has always been unduly magnified. The provision of track rails and travellers to which the guys of the ship can be attached, as is the practice in Germany, will tend to eliminate the source of trouble.

We must now consider the effect that weather will have on the big airship. In the past it has been a great handicap owing to the short hours of endurance, with the resulting probability of the ship having to land before the wind dropped and being wrecked in consequence. Bad weather will not endanger the big airship in flight, and its endurance will be such that, should it encounter bad weather, it will be able to wait for a lull to land. Meteorological forecasts have now reached a high state of efficiency, and it should be possible for ample warnings to be received of depressions to be met with during a voyage, and these will be avoided by the airship flying round them. In the northern hemisphere, depressions generally travel from west to east and invariably rotate in a counter-clockwise direction with the wind on the south side blowing from the west and on the north side blowing from the east. Going west, the airship would fly to the north of a depression to take advantage of the wind circulating round the edge, and going east the southern course would be taken.

Lastly, the vulnerability of the airship must be taken into account. Hydrogen is, as everyone knows, most highly inflammable when mixed with air. The public still feels uncomfortable misgivings at the close proximity of an immense volume of gas to a number of running engines. It may be said that the danger of disaster due to the gas catching fire is for peace flying to all intents and purposes negligible. At the risk of being thought hackneyed we must point out a fact which has appeared in every discussion of the kind, namely, that British airships flew during the war some 21 million miles, and there is only one case of an airship catching fire in the air. This was during a trial flight in a purely experimental ship, and the cause which was afterwards discovered has been completely eliminated.

For airships employed for military purposes this danger, due to the use of incendiary bullets, rockets and various other munitions evolved for their destruction, still exists.

Owing to its ceiling, rate of climb and speed, which we take to be from 70 to 80 miles per hour in the airship of the future, the airship may be regarded as comparatively safe against attack from the ordinary type of seaplane. The chief danger to be apprehended is attack from small scouting seaplanes, possessing great speed and the power to climb to a great height, or from aeroplanes launched from the decks of ships. If, however, the airship is fitted to carry several small scout aeroplanes of high efficiency in the manner described in the previous chapter, it will probably be able to defend itself sufficiently to enable it to climb to a great height and thus make good its escape.

The airship, moreover, will be more or less immune from such dangers if the non-inflamable gas, known as "C" gas, becomes sufficiently cheap to be used for inflating airships. In the past the expense of this gas has rendered its use absolutely prohibitive, but it is believed that it can be produced in the United States for such a figure as will make it compare favourably with hydrogen.

The navigation of an airship during these long voyages proposed will present no difficulty whatever. The airship, as opposed to the aeroplane, is reasonably steady in the air and the ordinary naval instruments can be used. In addition, "directional" wireless telegraphy will prove of immense assistance. The method at present in use is to call up simultaneously two land stations which, knowing their own distance apart, and reading the direction of the call, plot a triangle on a chart which fixes the position of the airship. This position is then transmitted by wireless to the airship. In the future the airship itself will carry its own directional apparatus, with which it will be able to judge the direction of a call received from a single land station and plot its own position on a chart.

We have so far confined our attention to the utilization of airships for transport of passengers, mails and goods, but there appear to be other fields of activity which can be exploited in times of peace. The photographic work carried out by aeroplanes during the war on the western front and in Syria and Mesopotamia has shown the value of aerial photography for map making and preliminary surveys of virgin country. Photography of broken country and vast tracks of forest can be much more easily undertaken from an airship than an aeroplane, on account of its power to hover for prolonged periods over any given area and its greater powers of endurance. For exploring the unmapped regions of the Amazon or the upper reaches of the Chinese rivers the airship offers unbounded facilities. Another scope suggested by the above is searching for pearl-oyster beds, sunken treasure, and assisting in salvage operations. Owing to the clearness of the water in tropical regions, objects can be located at a great depth when viewed from the air, and it is imagined that an airship will be of great assistance in searching for likely places. Sponges and coral are also obtained by diving, and here the airship's co-operation will be of value. Small ships such as the S.S. Zero would be ideal craft for these and similar operations.

The mine patrol, as maintained by airships during the war, encourages the opinion that a systematic search for icebergs in the northern Atlantic might be carried out by airships during certain months of the year. As is well known, icebergs are a source of great danger to shipping in these waters during the late spring and summer; if the situation becomes bad the main shipping routes are altered and a southerly course is taken which adds considerably to the length of the voyage. The proposal put forward is that during these months as continuous a patrol as possible should be carried out over these waters. The airship employed could be based in Newfoundland and the method of working would be very similar to anti-submarine patrol. Fixes could be obtained from D.F. stations and warnings issued by wireless telegraphy. Ice is chiefly found within five hundred miles of the coast of Newfoundland, so that this work would come within the scope of the N.S. airship. The knowledge that reliable information concerning the presence of ice will always be to hand would prove of inestimable value to the captains of Atlantic liners, and would also relieve the shipping companies and the public of great anxiety.

There are possibly many other uses to which airships can be put such as the policing of wide stretches of desert country as in Arabia and the Soudan. The merits of all of these will doubtless be considered in due course and there for the present we must leave them.

Finally, a few words must be written regarding the means to be adopted in introducing the airship into the realms of commerce. As we said at the beginning of the chapter it is not likely that the formation of a company to exploit airships only will at the present moment appeal to business men. Airships are very costly and are still in their infancy, which means that the premiums demanded for their insurance must of necessity be enormous. One suggestion is to place a reasonable scheme before the great shipping companies in case they will care to find the necessary capital and form subsidiary companies.

Another suggestion is that the Government should make arrangements to subsidize commercial airships. The subsidy might take the form of insuring them. If the burden of insurance is taken off their shoulders, it is considered feasible to promote companies which will give an adequate return for capital invested. The Government could also give a financial guarantee if mails are carried, in the same manner as is done by shipping companies.

In return for this the Government could at the outbreak of hostilities commandeer all or any of the airships for war purposes and so save the number to be kept in commission.

By this means the Government will have a large number of highly-trained and efficient personnel to call upon when the emergency arises, in the same way as the fleet can call upon the R.N.R. This system appears to be the best in every respect, and it cannot be denied that in the long run it would be the most economical for the country.

The airship has now arrived at the parting of the ways, and at this point we must leave it. The flying in war has been concluded, the flying in peace has not yet commenced. It seems a far cry to the dark days of 1914, when we only possessed two airships of utility, the one manufactured in France, the other in Germany, while to-day we have built the mighty airship which can fly to America and back. We are now at the dawn of a new period of reconstruction and progress, and during this period many wonderful things will happen. Not the least of these will be the development of the airship.

www.ingramcontent.com/pod-product-compliance
Lightning Source LLC
Chambersburg PA
CBHW031429210526
45464CB00005B/2112